T0222178

KARL POPPER

Sir Karl Popper was a major thinker of the twentieth century, one who – as Anthony O'Hear writes in his new Preface – 'has had a beneficent influence on those who have come under the spell of his thought and of the inimitable prose in which he articulates it'. It is now twenty-five years since Popper died, and thus seems – after a quarter of a century – an apposite moment to revaluate his impact, significance and influence. The several chapters in this classic volume focus on many key elements of Popper's thought and philosophy. They are by no means uncritical, but afford Popper the respect due to a philosopher who wrote always with a degree of clarity, precision and directness rare in the academic world of his time, and – as O'Hear puts it – 'even rarer subsequently'. This important book constitutes an essential introduction to some the most esteemed philosophical writing of our times.

ANTHONY O'HEAR, OBE, is Professor of Philosophy at the University of Buckingham. He is an honorary director of the Royal Institute of Philosophy and editor of the Institute's journal *Philosophy*.

TALKING PHILOSOPHY

General Editor: Edward Harcourt

The Royal Institute of Philosophy has been, from the very start, a fundamentally outward-facing organization. In 1924, Sydney Hooper – main mover behind the establishment of the Institute – realized that outreach to a wide interested public was a vital part of the value (whether social, cultural or intellectual) that philosophy at its best can impart. The Institute's first executive committee actively promoted that broad pedagogical message through accessible civic talks, and included in its ranks many of the most eminent luminaries of the day: not just professional philosophers but also sociologists, physicians, politicians, evolutionary biologists and psychologists. The Institute, from its foundation, has thus been rooted in an egalitarian community of people devoted to the principles of learning, debating and teaching philosophical knowledge in the broader service of what Hooper called 'the most permanent interests of the human spirit'. Talking Philosophy maintains this noble tradition. A book series published under the joint auspices of the Institute and Cambridge University Press, it addresses some of the most pertinent topics of the day so as to show how philosophy can shed new light on their interpretation, as well as public understanding of them.

Books in the series:

Moral Philosophy
The Philosophy of Mind
Ethics
A Philosophy of Need
Karl Popper
Spiritual Life

KARL
POPPER

EDITED BY

ANTHONY O'HEAR
University of Buckingham

CAMBRIDGE
UNIVERSITY PRESS

CAMBRIDGE
UNIVERSITY PRESS

Shaftesbury Road, Cambridge CB2 8EA, United Kingdom

One Liberty Plaza, 20th Floor, New York, NY 10006, USA

477 Williamstown Road, Port Melbourne, VIC 3207, Australia

314–321, 3rd Floor, Plot 3, Splendor Forum, Jasola District Centre,
New Delhi – 110025, India

103 Penang Road, #05-06/07, Visioncrest Commercial, Singapore 238467

Cambridge University Press is part of Cambridge University Press & Assessment,
a department of the University of Cambridge.

We share the University's mission to contribute to society through the pursuit of
education, learning and research at the highest international levels of excellence.

www.cambridge.org
Information on this title: www.cambridge.org/9781009230094

DOI: 10.1017/9781009230100

First published as Royal Institute of Philosophy Supplement: 39, Karl Popper:
Philosophy and Problems, 1995, Paperback ISBN 9780521558150.

This edition first published 2024.

A catalogue record for this publication is available from the British Library.

*A Cataloging-in-Publication data record for this book is available from the
Library of Congress*

ISBN 978-1-009-23009-4 Paperback

CONTENTS

CONTENTS

CONTRIBUTORS

Peter Clark

Donald Gillies

Peter Lipton

Graham MacDonald

Bryan Magee

David Miller

Kenneth Minogue

W. H. Newton-Smith

Anthony O'Hear

Michael Redhead

Michael Smithurst

Günter Wächtershäuser

John Watkins

John Worrall

E. G. Zahar

It is twenty-five years since Karl Popper died, and twenty-five years since the lecture series on which this book is based. In the original introduction to the book, I wrote that those interested in assessing Sir Karl's philosophical stature would find the essays contained therein a good place to start. This still seems to me to be the case. The essays focus on key elements of Popper's thought. They are not uncritical, but each of the contributors accords Popper a degree of respect and of respectful attention to his writing, and rightly so. For, whatever weaknesses there are in Popper's thought, he was and is a major thinker in the context of the twentieth century, and he has had a beneficent influence on those who have come under the spell of his thought and of the inimitable prose in which he articulates it. His style is all the more remarkable because he was writing in his second language, and always with a degree of clarity, precision and directness rare in the academic world of his time, and even rarer subsequently. Twenty-five years on, it also still seems to be the case that Popper had what I referred to in 1995 as 'a preternatural sense of where the deep issues lie' – again, something one would find it hard to discern in some of the most esteemed philosophical writing of the first decades of the twenty-first century.

In Popper's philosophy of science, the themes of induction, demarcation and progress dominate, topics which are taken up in a number of the essays in the book. It has to be said that these themes are not so prominent in the philosophy of science now as they were then, maybe because the questions Popper raised in these areas admit of no easy solution, at least in the terms in which he posed them. Moreover, Popper's thought in these areas had already been pretty exhaustively examined, even by 1995, at least in the terms in which he put the questions.

Pace Popper, induction in a broad sense is part of what our lives are based on. Learning from the past is part of what rationality consists in, and – again, pace Popper – it is hard to make sense of this without some inductive attitudes. Demarcation between science and other human activities, including morality and politics, remains a hot topic for a number of reasons, not least because scientists are increasingly brought into the political domain, and also because, for many, science is taken to be the touchstone of truth and rationality. What is not so clear is how helpful Popper's actual demarcation criterion is, subject as it has been to rather severe testing against actual scientific practice. Even so, there is still a lot to be said for his insistence on science's inherent fallibility – more than many are prepared to grant, especially when science is seen by so many as the benchmark of veracity. It is hard is to admit the fallibility of scientific theories (all our theories?); but it is even harder to grant fallibility while at the same time also recognizing the extraordinary progress science has made over the past three or four centuries. What remains crucial and valuable in Popper's approach to science, if not in the detail of his philosophy, is the insistence both on science's progress and on its fallibility – the way he tries to give due credit to each element of the situation. In Popper's thought, there is a pessimistic meta-induction, if you like – a sense that even the best theories at any time, including our own, will in the end almost inevitably be falsified; but he balances this pessimism with an optimistic assessment of science's propensity to progress – via the very critical spirit which leads us to recognize its fallibility.

Popper was well known for his hostility to the picture of science proposed by Thomas Kuhn in *The Structure of Scientific Revolutions*.[1] Indeed, in a paper entitled 'Normal Science and its Dangers',[2] Popper wrote that the picture Kuhn gives of 'normal science' (that is, when scientists train and work unquestioningly within a generally accepted theory or 'paradigm') is 'a danger to science and, indeed, to our civilization'. Popper does not deny that there is some truth is Kuhn's *description* of the scientific community, as it actually operates. In this picture, the community is at any one time dominated by some leading idea, and it is run by those propounding that idea in such a way as to crowd out dissent. With this goes all the institutional baggage we are familiar with in big science: funding, promotion, honours, initiation of students and researchers into the paradigm, peer review and all the rest of it. What Popper

---

[1] Thomas Kuhn, *The Structure of Scientific Revolutions* (Chicago: University of Chicago Press, 1962).

[2] Thomas Kuhn, 'Normal Science and its Dangers', in Imre Lakatos and Alan Musgrave (eds.), *Criticism and the Growth of Knowledge* (Cambridge: Cambridge University Press, 1970), 51–8, at p. 53.

denies is that this picture, true as it might be of the actual practice of science, can make room for the relentlessly critical activity he thinks science should be. Indeed, an institution of the type Kuhn describes will not be scientific in the way Popper wants. It will not encourage severe criticism of its own presuppositions. Hence, in Popper's view, the danger to civilization, given that he believes science and a properly scientific attitude is central to that civilization.

Can we say that both are right, on their own terms? What Kuhn describes is pretty much what goes on in the academic world in general – in science certainly, but also not unheard of in philosophy. This mode of existence may indeed be integral to any mature and smooth-running institution, including the institution of science, which would be Kuhn's point. Indeed, he argues that a subject not describable in his terms, and not dominated by a single ruling paradigm, would not *be* scientific: it would at best be pre-scientific, as he thinks psychology currently is. But if carried to the extreme, so that it is the whole story, a Kuhnian institution would suppress the spirit of criticism which Popper sees at work in the moments of science he so admires, and what for him makes science so central to our civilization. The key example of such a moment for Popper was, as he tells us, when, after two or more centuries of pretty well universal acceptance in the scientific community, Newton's theory was refuted by Einstein's. Popper would not, I think, be impressed to be told that 97 per cent of scientific publications say so and so, as if that were an argument in their favour. For Popper, if not for Kuhn, that would be quite the wrong sort of argument, and quite the wrong sort of attitude. It is an attitude which is inherently suspicious of the type of openness which Popper thinks should characterize good science, where even (or, perhaps, especially) the best theories are continually subjected to the severest criticisms and thus tested in confrontation with competing theories.

And this brings us to openness more generally. For Popper sees his ideal open society as the scientific method – or what he conceives as the scientific method – writ large, ruling over society as a whole. An open society is one in which laws, policies and institutions are subjected to continuous criticism from any quarter, particularly from the least privileged, from the people most likely to be adversely affected by any policy or institutional set-up. An open society will not be ham-strung by taboos or uncriticizable traditions, and its members will be regarded as individuals, rather than as members of groups or as determined by their birth or the traditions into which they are born. Even in initially expounding a vision of what he calls the abstract society, Popper admits that there is a degree of utopianism in his picture: there can never be 'a completely abstract or even a predominantly abstract society', nor can there ever be a completely rational society. Our emotional needs are such that we need social and spiritual bonds of various sorts, beyond the rational-critical.[3] And, later on, Popper came to recognize that the type of liberal, open society he

---

[3] See Karl Popper, *The Open Society and its Enemies* (5th edition) (Abingdon: Routledge and Kegan Paul, 1966), Vol. I, 175.

advocated itself needed traditions of liberal openness to guide its members and to pre-dispose them to liberal attitudes.

However, none of these qualifications, necessary as they may be, ever deflected Popper from his conviction that a decent society will treat its members primarily as individuals, rather than as members of groups or collectives – and also that in such a society anyone should have a right to be heard, whatever his or her origin or status in that society. His anti-collectivism went in hand with his admirable repugnance for those theories of history which saw history as moving in a given direction, with those unsuited to that movement being dispensable and disposable. History has no determinate direction, and its outcomes are unpredictable. It is not just immoral to sacrifice today's individuals for some future utopia or goal; it is also epistemologically unsound, because we cannot predict the future course of history. And one key reason for this epistemological incapacity is that we cannot know now what scientific or technological developments there will be. If we knew now what would be available in, say, twenty-five years, we could do it now – which, of course, we can't. When this volume came out twenty-five years ago, we had no clue as to the way the Internet and all its associated paraphernalia would have developed by 2020 (and probably a less than complete understanding even now of what it all means or amounts to).

It seems to me that, given the way that identity politics and group thinking have developed in recent years, if this current volume were to be produced now, more might be said about the implications of Popper's anti-collectivist views, and also about his insistence that the source of an opinion should always be distinguished from its validity. A valid opinion is valid, whatever its source; and all opinions should be judged on their merits, regardless of their source or the ethnicity or social standing of their proponent. And disagreement, even if unpalatable, should be welcomed, as a key part of the critical spirit. Further, while he was as attuned as anyone to environmental threats and dangers (particularly those arising from the disposal of nuclear waste), Popper would also have something to say about the hazards involved in impoverishing people living now in the light of futures half a century ahead, especially given that we cannot know now what technological developments there will be over the next half-century. And, more generally, he would have been more alive than perhaps many were in 1994 to rhetoric about the need for change and the obligation to prepare for the future, which is usually rhetoric aimed at promoting things the speaker is keen on. But the speaker knows no more than any of us about what the future will bring, or about the paths changes will take and make. In speaking of the need to prepare for change, most of us are too ready to assume we know what will come about. This is not a mistake a Popperian should make – and not just because of his or her general fallibilism and doubts about inductivism. It is also a feeling ingrained in Popper's humane approach to social questions, as manifested in his writings on the open society.

Looking back at where Popper stood twenty-five years ago, and might stand in the future, we might conclude that his philosophy of science is well established as one of the landmarks in the subject's development. But, because it is so well established and so well known, it has somewhat fallen out of the current limelight. His views on society and politics are less well developed and their implications less well explored. But, as they were important and influential in the battles against communism and totalitarianism in the 1940s and 1950s – and bravely so in the climate of those times – so they have relevance in the issues confronting us in the 2020s. They deserve consideration and study, and more consideration and study than we gave them twenty-five years ago.

Introduction

ANTHONY O'HEAR

This collection of essays on the work of Sir Karl Popper is based on the Royal Institute of Philosophy's annual lecture series given in London from October 1994 to March 1995. Popper himself died in August 1994, shortly before the start of the lectures. His death was the cause of sadness to all of those involved in the series. Some, indeed, had been close friends of Popper over many years, and others colleagues and acquaintances, some close, some more distant. Even those unacquainted with Popper personally spoke in their lectures of the profound intellectual stimulation they had received from the study of his works.

Towards the end of the course of planning the lecture series, I did, with some trepidation, contact Popper. His reaction was at once generous and self-effacing. Having initially told me that he did not envy me my task of getting speakers, when he saw the outline programme, he wrote that 'the plans for the course on my philosophy were very interesting: much more interesting than I thought possible'. Credit here should be given where it is really due. Once the Royal Institute determined on the topic, both subjects and speakers suggested themselves naturally; and there was no difficulty in persuading potential contributors from

Britain to participate. Popper himself suggested that Günter Wächtershäuser from Munich and Hubert Kiesewetter from Eichstatt should be added to the original list of British-based contributors, and this was done. Thanks are due to all who took part in the series, and who have helped to make this book as comprehensive and wide-ranging as it is.

Popper also agreed to take part in a question and answer session at the end of the series. That this was not to be is a matter for great regret, both personally and intellectually. Many original observations and criticisms were made during the course of the lectures, to which it would have been fascinating to hear Popper's own reactions. Despite the sense of regret shared by all involved in the series, however, the lectures as given and as here reproduced engaged fully and critically with Popper's philosophy. They are neither eulogistic nor valedictory, but testify to the sense that, whether Popper himself is alive or dead, his ideas continue to pose problems, to have consequences still to be fully explored and so to bear intellectual fruit.

Popper's philosophy is marked by a breadth and a coherence unusual for a modern philosopher. While his fundamental insights may stem from the philosophy of science, what he has to say there reaches out into politics, into the theory of rationality and into the nature of life itself. On science, Popper's thought is marked by a deep hostility to any profession of certainty or to any claim to justification. He accepts Humean scepticism about induction, taking on board the consequence that this means that we can never know whether any universal theory is true. He believes that even observation statements implicitly use universal theories because in referring to objects such as glass or water we are making claims about how they will behave in the as-yet-unknown future. His scepticism thus runs deep, but he thinks he can base an account of scientific rationality on the negative activity of attempting to disprove theories. The empirical disproof of a theory is conclusive, while any amount of evidence in favour of a theory remains inconclusive. True scientists make bold conjectures and then, equally boldly, attempt to refute their conjectures by the severest tests they can devise. Following this procedure, we are entitled to accept as yet unfalsified theories provisionally, though we should not think this means they are in any sense justified. True science, indeed, is demarcated from other activities by the rigorous acceptance of the method of falsification and its results.

While some unscientific activities, such as Marxism and psychoanalysis are intellectually disreputable in that they pose as science while refusing to accept empirical disproofs as conclusive, there are intellectually respectable pursuits which are not scientific. Examples would be mathematics, ethics and philosophy itself. While not being susceptible to empirical disproof, they do all have well-established traditions of criticism to underpin their rationality. Rationality is thus seen by Popper as the generalized application of the critical method.

In science Popper's stress on criticism combined with a strong commitment to realism leads him to develop an original line on probability. He regards probability statements as objective and falsifiable. They are not to be seen as expressions of our ignorance about the full causal determinants of events, but as describing actual, but non-deterministic propensities in the real world. The world is not wholly deterministic, but in many areas is governed by such propensities producing real but non-determining tendencies for events to happen in particular ways. Popper's commitment to indeterminism is closely linked to belief in human freedom and creativity, which he believes would be ruled out by any form of determinism. His belief in propensities allows him to think of probabilities as objective forces which leave room for the exercise of freedom.

Criticism, freedom and rationality are central to Popper's views on politics and open society, views which have struck a resounding chord with those in Eastern Europe and elsewhere who have lived and suffered under dictatorships. As an aspect of Popper's general epistemological scepticism and his hostility to justificationism in any form, we are told that any of our actions and policies are likely to have unforeseen and unintended consequences. This is significant particularly where large scale political changes are being attempted. We should, therefore, be suspicious of rulers and politicians who wish—even for the best of motives—to impose comprehensive blueprints on a society. Far from acceding to such dictatorial ambitions, we should work for open societies, ones in which anyone is entitled or even encouraged to criticize a policy, and in which rulers can be removed by the ruled regularly and peacefully. Accepting their own inevitable ignorance of the effects of policies, rulers should confine their activities to the eradication of manifest evils, rather than attempting to impose their untried and possibly unwelcome visions of happiness on the rest of the population.

Science and politics, then are ideally to be characterized by an admission of our ignorance and by the attempt to weed out false theories and to remedy the negative effects of our policies. Life itself comes to be seen by Popper in very similar 'problem-solving' terms. In the process of evolution, all kinds of modifications to existing creatures occur. Like a false scientific theory, most of these modifications are ruled out by the refuting environment. It is the environment, indeed, kicking back which assures us that our theories are about a real world and, in various directions, making progress. But in our scientific theorizing we are following the same evolutionary sequences as the most primitive amoeba, going from initial problem to an attempt to solve it. Then after eliminating error from the proposed solution, with luck we may reach a partial solution, and so move on to new problems. The difference between human beings and other life-forms is that we can make our modifications and propose our attempts at solution exosomatically, in symbols, outside our bodies. The criticizing environment can attack our theories, which die in our stead, rather than, as in the case of biological evolution, the modified organism.

The main lines of Popper's thought are clear, comprehensive and far-reaching. As would be expected, his doctrines brought him into conflict with many of the intellectual fashions of his and our times—with, for example, attempts to work out positive theories of induction, with anti-realism in the philosophy of science, with subjectivism in quantum theory, with Marxism in politics and with deterministic accounts of history and of our future. All these disputes and controversies are alluded to and pursued in the essays which follow.

The first four essays deal with some familiar and basic problems arising from Popper's account of science. W. H. Newton-Smith is the latest in a long line of critics of Popper who wonder whether he can really be said to have dispensed with induction in his account of science. Newton-Smith, though, takes the criticisms a stage further than most, suggesting that philosophy of science should abandon the attempt directly to defend a particular method for science whether falsificationist, inductive of some other. What it should do is to analyse scientific rationality in terms of that institutional values, though whether one is thereby entitled to regard them as justified or desirable because of the success of science is less clear.

Peter Lipton also considers Popper's anti-inductivism, and contrasts it with what he calls a reliabilist approach to knowledge. On this approach one may be said to know something if one has acquired a true belief through a method which is in fact reliable. One does not have, over and above actual reliability, to prove that the method in question is bound to work. If inductive methods (or some inductive methods) are in fact reliable, contra Popper, they may thus be said to produce knowledge. Philosophical opinion will be divided as to whether this essentially naturalistic approach to knowledge is an advance in epistemology or its final abrogation, in that the approach starts by assuming that we can in fact identify bundles of true theories. Lipton, though, goes on to argue that a Popperian method of falsification may be not just necessary for positive knowledge, in that positive claims to knowledge have to survive attempted falsification, but also sufficient for it. This is because there can, in fact, be no falsification without a background of accepted truth, which is an interesting way of looking at the familiar suggestion that Popper's method of falsification needs some basis of justified truth on which to stand.

With that, Elie Zahar would agree, though he conceives the basis rather differently from Lipton. Zahar accepts Popperian scepticism about general theories and even about singular observation statements where that is observed are held to be objects and states of affairs in the external world. But, following Brentanco, Zahar makes out a strong case for regarding statements about one's current psychological states as both justified and—perhaps more controversially —as the explanada of the theories of science. According to Zahar, then, we should regard such statements as the firm and justified empirical basis for science, something Popper would emphatically reject, but without some such basis, his system has seemed to critics to be hopelessly drifting in the shifting currents of theory.

Taking up some of the controversies between Popper and Kuhn, Lakatos and other philosophers who have examined the history of science, John Worrall argues that Popper's view of scientific theory is over-simplified. In particular, innovations in scientific theory should not be seen, as Popper would wish, as bold, imaginative conjectures produced, like Darwinian mutations, without any instruction from without. Worrall shows how, in the case of Fresnel's development of the classical wave theory of light, the new theories were produced in a thoroughly directed way, by systematic and logical argument from what was previously known. While not strictly incompatible with the broad lines of Popper's falsification, Worrall certainly does something to qualify Popper's more extreme rhetoric about the utterly un-Baconian nature of the scientific process.

The next three papers all focus on Popper's propensity theory of probability and his commitment to indeterminism. Donald Gillies suggests that the propensity theory fails to solve the problem of the objectivity of singular probability statements—that for which it was originally proposed—but argues that the theory is nonetheless desirable in avoiding the operationalism inherent in the frequency theory. He goes on to sketch a broadly Popperian account of the falsification of probability statements, and ends by showing that corroboration is not a probability function; Popper was anxious to defend this view as part of his anti-inductivism, though, as he says, Gillies arrives at the same conclusion by a distinctly un-Popperian route.

David Miller accepts that determinism, as a philosophical thesis, is empirically unfalsifiable (and hence, in Popper's terms, 'metaphysical'). However, various difficulties with completing deterministic accounts of the physical world ('scientific determinism') do expose its status as a metaphysical rather than a scientific theory and ought to lead to its rejection. Miller goes on to examine one of Popper's favourite arguments for indeterminism, that from Landé's blade, and finds it inconclusive in that regard. In the final part of his paper, he shows that Popper's most recent account of propensity, as that which allows genuinely new possibilities to emerge, is something over and above what is meant by probability in the probability calculus; for in that calculus genuinely new possibilities must have zero probability. Miller, nonetheless, concludes with the claim that zero propensity need not imply impossibility, and that many things which have actually happened, such as the painting of the *Night Watch* or the building of the Parthenon, cannot have been foreshadowed by propensities in the first few seconds of the universe.

That the link between Popper's views on indeterminism and his belief in genuine creativity is by no means clear is the conclusion of Peter Clark's paper. Clark admires the seriousness of Popper's commitment to human freedom and creativity, but questions the relevance of Popper's arguments about unpredictability to this. After all, things, including ourselves, may be unpredictable without being undetermined. Clark accepts that the propensity theory is a bold attempt to solve the thorny problem of the existence of stable, non-trivial statistical regularities in the physical world, but he is dubious that it solves other problems to which Popper has applied it, such as the measure zero problem in statistical mechanics or the paradoxes in quantum theory.

In the first of two papers on the application of Popper's thought to specific areas of science, Michael Redhead focuses on Popper's incursions into quantum theory. These go back as far as 1934 and continued well into the 1980s. While, as Redhead shows, Popper's suggestions are flawed in detail, it is certainly arguable that Popper's 1934 article may have influenced Einstein in what has come to be known as the EPR paradox (published in 1935). Popper's interest in quantum theory was from that start motivated by a strong commitment to realism, in an area and at a time when realism was distinctly unfashionable. That in itself put Popper in the Einstein camp, though his later espousal of indeterminism took him partly outside it. Apart from the details given by Redhead (much deriving from private conversations and correspondence, and hitherto unpublished), what is fascinating about Popper's dealings with quantum theory is that in them we see the method of conjecture and refutation at work in practice, as well as a readiness on Popper's part to bow to criticism and refutation which has not always been so evident in his pure philosophy.

Günter Wächtershäuser also relates Popper's philosophy of science to a specific area of science, in this case to biology and the study of the origin of life. In considering the work of van Helmont, Berselius and other pioneers in the field, Wächtershäuser argues that biology achieved the Popperian goal of moving ever closer to the truth by definitely non-inductive methods. More recent work, however, on the pre-biotic broth from which life is supposed to have originated, has been dominated by inductivism, and, in Wächtershäuser's view, has been largely inconclusive. Wächtershäuser concludes his paper by outlining his own original theory on the origin of life. Though initially presented without observational or experimental support, it has succeeded in capturing the attention of many scientists by virtue of its explanatory power. Work is currently underway to test and refine the theory empirically.

In the first of two papers on Popper's discussion of the theory of evolution, John Watkins develops a refinement of that theory originally suggested by Popper himself in 1961. Dubbed by Watkins the 'spearhead model', the suggestion is that in complex organism there are control systems as well as the physical hardware of limbs, organs and other bodily parts. In evolution the two systems may develop independently. The control system may outreach the motor system in ambition, and thus be able to exploit changes in the motor system, which would otherwise remain unused even though potentially advantageous. Evolutionary development is thus led by advances in control systems. We can thus see why even small structural changes can be useful to organisms, and also how certain rather larger mutations can be directed to useful purposes, rather than remaining unused or even awkwardly obstructive. Watkins thinks of consciousness in human beings as our control system; he also suggests that the impetus of sexual selection may well drive our control system to develop capabilities in us, like the concert pianist's ability to play Chopin, which are well above what is necessary for our survival.

Evolutionary over-reaching also concerns Michael Smithurst, and in rather similar terms. After echoing some of Worrall's reservations about Popper's Darwinian analysis of scientific theory formulation, he goes on to consider what Darwinism might say to such traits as the ability to do higher mathematics or astrophysics. Smithurst's answer is that, while evolution in the struggle for survival in the physical environment might have fitted us to crack walnuts, it has little direct bearing on many of our more flamboyant activities, intellectual and cultural. These have developed both because of our neoteny—which means that human beings are born more immature and hence mentally more malleable than many other creatures—and also because of our propensity to engage in tournaments of the mind to attract mates, which is part of what sexual selection amounts to in the human case. While none of this can be a complete account of what is going on when we research into the quantum world or enjoy the painting of Turner, it is certainly a useful corrective to those theorists of evolution who tend to epistemological scepticism by seeing ourselves and our faculties as limited and conditioned by our immediate physical environment.

There follow three papers on Popper's social and political philosophy. Kenneth Minogue takes issue with Popper's analysis of social and political life in terms of problem-solving, and also with his proposal that essentially the same methods are to be used in the physical and social sciences. Minogue argues that in the description and explanation of human actions and in historical narration generally what is at issue is not the attempt to subsume particular events under general laws (which is Popper's model of explanation in both the human and the physical worlds). In even the most basic description of human action, there is already an element of explanation: we always describe actions in terms of the reasons and intentions we suppose their agents to have. Moreover, this assumption that even at the most basic level human action is seen in terms of reason giving and reason seeking is always embedded in the unspoken patchwork of habit, routine and principle within which we live our lives and against which some, but not all situations present themselves as problems to be solved. Minogue does not have much to say in answer to the difficult problem which arises when agents from one habitual background confront those from another, but Popper gives little help here either. His own discussions of value, thought at one level unexceptional and deeply felt are, as Minogue avers, analytically rather thin.

Graham Macdonald engages in a careful examination of the tangled web of anti-historicist arguments in Popper's *Poverty of Historicism*. He is not convinced that Popper's insistence on the role of unpredictable development in human knowledge leaves the would-be predictor of the future course of history without recourse. Important, though imprecise lines of development might still be discernible and useful in making predictions. Macdonald considers G. A. Cohen's work on technological development and Jack Goldstone's 'political stress indicator' as pointing to tendencies on which predictions might possibly be based, though whether Popper would regard this as an interestingly strong version of historicism is open to doubt. (Popper does not, after all, rule out all talk of trends in history, so long as we do not regard trends as imprisoning agents in specific and inflexible futures). Macdonald also examines Popper's advocacy of small-scale changes in politics. He points to some of the difficulties in deciding just what might or might not be regarded as small-scale, makes the interesting observations that on some occasions what was intended by policymakers as small-scale changes has far-reaching and even revolutionary results.

Bryan Magee has had both academic and political experience. In his view, too much political discussion, particularly among intelligent people, is over-theoretical, tending to produce blueprints divorced from what actually exists and from people's actual motives and aspirations (which are continually changing anyway). Magee advocates the Popperian approach to politics, one which treats political activity as essentially a matter of finding solutions to genuine problems, and which works through discussion and criticism, and a willingness to listen to one's opponents. Popper's political views are certainly in the liberal-democratic tradition, and though Popper (like Magee) had little time for much of the cultural pessimism so fashionable among intellectuals, Magee is at pains to emphasize that Popper is not a conservative (in the sense that Minogue's paper, for example, is). Popper's attitude to institutions is radical, rather than conservative, requiring (like Baroness Thatcher) that they be continually and at times disruptively subject to scrutiny and criticism. Though he is convinced of the importance and vitality of Popper's political philosophy, however, Magee concludes his paper with an example of what Popper himself might be accused of having taken an over-intellectual attitude to a political issue; even the arch-critic of intellectualism in politics could not suppress all taint of his métier.

What will be the estimate of Popper in 200 years time? Of course, it is too early to judge that. What certainly is true is that at the end of the twentieth century, Karl Popper does appear to be one of the more considerable philosophical figures of the age. That is not to say that he has ever exactly been fashionable (though it has to be to said that his frequently expressed claim to have been neglected by academic colleagues is exaggerated and, in retrospect, looks ill-judged). But Popper has had no part in linguistic philosophy in either its ordinary language or its Davidsonian phases. His hostility to justificationism has not led him to develop an interest in what is known as naturalized epistemology, or in Wittgensteinian approaches. He had little time for the Carnapian fashion for formal definition which was for a time dominant in American philosophy of science; the minutiae of the more recent realist–anti-realist debates on both sides of the Atlantic have largely passed him by, as have the intricacies of possible worlds logics and so-called cognitive science. Despite having a considerable interest in the self and the mind, he has showed little interest in contemporary philosophy of mind, and much the same could be said of his relationship to the contemporary political philosophy of Rawls and Nozick.

It is not, of course, that from his philosophical standpoint Popper would not have had things of interest and relevance to say on any of these topics, and on many more. It is rather that he has shown a determined independence of mind on all the topics in which he has had an interest. According to his friends, this independence of mind has enabled him to avoid wasting his time on transient fads, while his critics will see its effect in terms of a systematic failure to engage with philosophy at its leading edge.

Not to be moved by the spirit of the time or, rather, not to be influenced by rhetoric which relies on such persuasive definition, is entirely appropriate for a leading critic of historicist notions. And it is in this area, and in the area of politics, social science and psychology more general that Popper's influence has been most widespread and most salutary. On his death, more than one newspaper obituary was headed 'The Man Who Killed Marx and Freud'. Academics may turn their noses up at such sub-editorial crudity, and they will doubtless also argue that Popper's criticisms of Marx and Freud are not themselves irrefutable. It is, indeed, the nature of intellectual endeavour to be but rarely conclusive.

What, though, is not open to doubt, and why I for one will always be grateful to Popper, is that during some very bad times intellectually, he was hugely influential in creating a climate of opinion in which it became possible to hold that there might be something deeply amiss with both Marxism and psycho-analysis on methodological grounds. And he did this while having no truck with irrationalism or with what he called oracular philosophy. In like manner, his staunch defences of realism in the area of quantum physics and probability theory, though once again deeply unfashionable at the outset, have provided timely and salutary reminders that physics need not devolve into irrationalism, and should not. More generally, as several contributors to this book testify, there is much which is both challenging and moving in Popper's advocacy of openness in politics and his defence of creativity in human nature. In a century which has been peculiarly susceptible to irrationalism and inhumane dogma, Popper's powerful defences of individualism and of reason at times take on an inspiring aspect.

On the other hand, when we come to look at the detail of Popper's arguments, the picture is less clear. Few qualified observers believe that Popper has succeeded in solving the problem of induction or in presenting a non-inductive account of science, and many find his model of scientific theorizing over-simplified. Considerable doubts remain about the detail and even the significance of the propensity theory of probability. It also remains unclear how far his account of explanation in the social sciences is acceptable, or to what extent a Darwinian model is appropriate to human thought. In all these, and in other areas, the question is not so much the rightness or wrongness of Popper's views, nor is there any doubt that unlike many contemporary philosophers, Popper has a preternatural sense of where the deep issues really lie. The question we must consider is the extent to which the arguments he advances actually lead us to a deeper understanding of the issues involved. Those interested in coming to a provisional estimate of Popper's philosophical stature could do much worse than read the essays which follow with this question in mind.

1 Popper, Science and Rationality

W. H. NEWTON-SMITH

We all think that science is special. Its products—its technological spin-off—dominate our lives which are thereby sometimes enriched and sometimes impoverished but always affected. Even the most outlandish critics of science such as Feyerabend implicitly recognize its success. Feyerabend told us that science was a con-game. Scientists had so successfully hood-winked us into adopting its ideology that other equally legitimate forms of activity—alchemy, witchcraft and magic—lost out. He conjured up a vision of much enriched lives if only we could free ourselves from the domination of the 'one true ideology' of science just as our ancestors freed us from the domination of the Church. But he told us these things in Switzerland and in California happily commuting between them in that most ubiquitous product of science—the aeroplane.

We are so impressed in general with the results of science that we give quite amazing credibility to any claim that is successfully represented as being the result of a scientific investigation. As an illustration one might cite the credibility given by the general educated population to claims about the origin of the universe notwithstanding the paucity of evidence and the fact that cosmological theories have a half-life of ten years or so. But my favourite example is much more mundane: it is the Ben Johnson fiasco. The entire Canadian population was convulsed with delight the night Johnson appeared to have won the gold medal in the 100 metres. The Prime Minister had declared him a new model for the youth and a symbol around which to build the Canadian identity. However, at a press conference the next day it was revealed that a test which was probably understood by only a handful of Canadians, applied to data available only to a couple of scientists, had shown he was on steroids. Notwithstanding the nation's desire to believe the contrary, within hours everyone concluded that he had cheated: such is the power of science. No other institution in western pluralistic societies has that power to engender belief.

In view of the success and the status of science it is not surprising that in this century, the century of science par excellence, the century of the vast majority of scientists who have ever lived, philosophers of science have been preoccupied with trying to understand what it is that makes science successful. However strongly we may feel this drive to understand, it is not hard to imagine that it was even stronger for the young Popper in the Vienna of 1919 when the success of Einstein's Special Theory of Relativity, followed by the successful test of the General Theory by Eddington, vindicated the claims of the new science. In various places, Popper graphically described the excitement his circle found in Einstein's theory and his growing dissatisfaction with Freud, Adler and Marx. He thought he had found the answer to the question of what makes science special in his principle of demarcation: 'the criterion of the scientific status of a theory is its falsifiability'. Einstein's theory led to a prediction. When Eddington set off to make the appropriate observations the theory was put at risk in the eyes of the scientific community. Things might not have turned out as predicted, in which case Einstein's theory would not have been in trouble. But according to Popper, neither Freud nor Adler nor Marx held theories at risk from testing.

An important qualification was necessary as Popper came to realize. There are two ways one can fail to be scientific. One can hold a body of propositions none of which leads to any prediction whatsoever. Or one can hold a body of propositions which do lead to predictions but hold those propositions dogmatically. In this case one refuses to put the theory at risk by invoking special pleading when things go wrong. But whichever way we look at the psycho-analysts and the Marxists they certainly were not having the success of Einstein.

Popper's project at the time was an instance of an approach which I once called the construction of **rational models of science**. The underlying idea is the plausible assumption that the success of science derives from the use of the scientific method which is characterized by the model.

A method is always a method to some end. And so in constructing such a model, one has to specify the aim of science, usually some noble (or perhaps it is nobel) aim. To vindicate such a model one needs some kind of argument to show that adherence to the method as characterized leads to the goal as characterized.

To explain the special success of science we would need one further ingredient. We have to establish that scientists, members of the scientific community, have specially committed themselves to the goal and the method. That is, they bind themselves to following its dictates regardless of conflicts with personal interest (by and large). Of course exceptions can be allowed without adversely affecting the viability of the model. It has been alleged that Cyril Burt's desire to 'establish' the greater importance of genetic factors over environmental factors led him to invent the data and indeed to invent research assistants to explain how he was able to process the data in question.[1] But so long as this sort of example represents no more than the occasional aberration, it does not affect the viability of this rational model approach to explaining the success of science. I called this the **rational model approach** because it was meant to capture the thought that science is the paradigm of institutionalized rationality; that science had some special unique form of rationality embodied in the methods of science which scientists heroically followed.

More recently I have become dubious about this approach to science in spite of its initial plausibility. Someone else who shares this scepticism is Philip Kitcher who has graphically described the second aspect of the approach in what he calls the myth of **Legend**.

> Legend celebrated science. Depicting the sciences as directed at noble goals, it maintained that those goals

---

[1] Since the late Seventies, these allegations were generally accepted as being well-founded. More recently this has been challenged. See Robert B. Joynson, *The Burt Affair* (London: Routledge, 1989).

> have been ever more successfully realized. For
> explanations of the successes, we need look no further
> than the exemplary intellectual and moral qualities of the
> heroes of Legend, the great contributors to the great
> advances. Legend celebrated scientists, as well as science.[2]

On the rational model approach or in the myth of Legend we explain what is special about science and scientific success by reference to a characterization of the goals and methods of science and by reference to the heroic virtues of individual scientists.

There is a vast array of rational models on offer. They vary in their accounts of the goal of science: truth; explanatory power; increasing verisimilitude (that is, roughly speaking increasing approximation to the truth); or simply predictive and manipulative power at the observational level. It is curious that there is so much discussion of what something so important to us as science is actually aiming at. In order to progress without too much detouring I am going to distinguish between the **manifest aim** of science and the **latent aim**. By the manifest aim I mean that phenomenon with which I began—predictive and manipulative power at the observational level (what Bacon had in mind by 'power over nature'). There is general agreement that science achieves this. For some philosophers, the instrumentalists, that is all there is. For others, particularly the realists, progress towards the manifest aim is a sign, albeit a fallible sign, of progress towards the latent aim of science

---

[2] P. Kitcher, *The Advancement of Science* (Oxford University Press, 1993), p. 3.

which might be Truth or, with Popper, the more modest one of increasing verisimilitude.

Equally there is great variation in the characterizations we are offered of the scientific method. We have Mill's methods, the inductive logic of the positivists, Popper's falsificationism and contemporary Bayesian accounts among others. Notwithstanding these rivalries, virtually all philosophers of science assumed that an enlightening linguistic characterization can be given of the principles which should guide us in choosing between hypotheses and theories. In so far as those who have sought such accounts recognized difficulties they were seen as mere barnacles on the ship of progress which would be removed in due course. Thus Ernest Nagel referred to our 'basic trouble in this area of inquiry' as being that:

> we do not possess at present a generally accepted, explicitly formulated, and fully comprehensive schema for weighing the evidence for any arbitrarily given hypothesis so that the logical worth of alternative conclusions relative to the evidence available for each can be compared.[3]

I no longer think that there is a characterization of the goals and methods of science which, together with assumptions about the virtues of scientists, will fully explain the success of science. In indicating the reasons for my dissatisfaction I am

[3] E. Nagel, 'The Logic of Historical Analysis', in H. Feigl and M. Brodbeck (eds), *Readings in the Philosophy of Science* (New York: Appleton-Century-Crofts, 1953), p. 700.

going to do three things. First, I will look at Popper's theory of science as a candidate for the rational model of science. I will argue in familiar ways that the grand Popperian experiment of falsificationism fails. However, I hasten to add, that the other non-falsificationist players in this particular game also fail. And I will argue that a movement in Popper's thinking indicates that he saw one of the fundamental problems with the rational model approach. In a nutshell the problem is that all characterizations offered of scientific method at the level of generalization and abstraction favoured by philosophers of science fail to be an account of anything specifically scientific. Hence such stories cannot account for what is special about science. What we end up with are at best characterizations of general epistemic virtues and not specifically scientific ones. I will then indicate the direction in which we should look if we are seriously interested in finding what is special about science. I will argue that Popper himself clearly and insightfully indicated that this is the direction one should go. It is a direction which introduces a social element into our story. This is an approach that has received little attention by philosophers of science in general and by Popperian philosophers of science in particular.

Popper's theory of science, were it successful, might appear to answer our question by explaining what it is that makes science so successful and what thereby legitimates the epistemic status we attach to scientific results. That theory of conjectures and refutations represents science or good science as a matter of making bold conjectures which are subjected to rigorous testing. The bolder the conjecture, the more it rules out, the better. And the more rigorous the testing, the better. This is the scientific method and adherence to that method is what makes science the success it is.

This framework did not rest for Popper simply on a generalization (heaven forbid) from cases antecedently recognized as science and cases antecedently recognized as pseudo-science. It rested ultimately on an *a priori* philosophical argument of Hume. Hume was entirely happy with deductive arguments. Any argument in which the premises entailed the conclusion was fine by him. Arguments in which the premises could be true but the conclusion false had no rational force for him. This is not what we prephilosophically think. We think that there can be good arguments, inductive arguments, in which the premises support the conclusion without entailing it. Unlike Hume, we think the fact that unadulterated bread has always nourished us in the past provides a decent reason for thinking that it is at least likely that tomorrow's bread will also nourish us. Hume denied this. He held that any attempt to justify accepting a conclusion not entailed by the premises would run into vicious circularity. Induction does not have the sanction of reason.

I draw attention to this familiar point about Hume to highlight something that is not always noticed by all readers of Popper. A careless reader might think that Popper is simply rejecting certain kinds of inductive arguments such as induction by simple enumeration. But Popper takes Hume seriously and that means rejecting **all** non-deductive argument. Galileo posited the existence of mountains on the moon in order to explain the changing patterns he observes on the moon's surface. For Hume, Galileo had no reason whatsoever for this conclusion. Einstein argued against Mach that one should believe it true that there are atoms because positing the existence of atoms would explain Brownian motion, but he does not have reason on his side. Indeed, he has no reason to think it true or even to think it reasonable to believe that probably there are atoms. This strategy of coming to believe in the truth or the probable partial truth of hypotheses by reference to their explanatory power (called **inference to the best explanation**) is a particular species of induction.

So too are certain exercises of judgment. Knowing that I have had a successful track record in identifying edible fungi, I conclude on the basis of certain characteristics that a particular fungus is edible. I trust my judgment. I believe that I have good but fallible reasons for thinking that this mushroom is going to be edible. It is often said that Hume's arguments may worry me in tutorials but not in the fields, and that I am an inductive creature, but Popper will have none of it.

I am going to call this anti-inductive aspect of Popper's account of science his **Grand Experiment**. By that I mean simply the attempt to construct a theory of science in which induction plays no role whatsoever. Notoriously Hume thought that we had no choice but to proceed inductively. It was part of our nature. For him it was a **bemusing** feature of the human condition that we as philosophers could show that what we as humans do does not have the sanction of rationality. For Popper, scientists neither do nor should indulge in induction.

If a particular theory of science is to provide a rational model which will be part of the explanation of the success of science, we need to be able to show that there has been progress towards the goal as characterized and that following the scientific method as characterized has played a crucial role in making progress. How does Popper fare on this approach? Our best theories in the past have turned out to be falsified. Popper expects no less of our current theories. If one expects theories to turn out to be false, it would hardly be rational to make the goal of science the discovery of true theories. So Popper, as noted above, offers an apparently more modest goal. This is to have theories of increasing verisimilitude; that is, theories which either contain more truth and no more falsity than their predecessors, or theories which, without decreasing the truth they contain, decrease their falsity content.

If that is the goal of science, is there any reason to think that science progresses? There is an argument to this conclusion which I and other philosophers of science once thought had some power. It begins by noting that contemporary theories such as Einstein's are much more predictively successful than their predecessor theories such as Newton's. It was argued that it would be very surprising that theories could be more successful in this way unless they contained more truth at the theoretical level. That is, the best explanation of their predictive success lies in the assumption that we are getting more things right at the theoretical level. Our theories getting things right at the observable level is explained by their getting things right at the theoretical level by uncovering underlying structures in the world. But this argument is a grand inductive argument. It is a case of inference to the best explanation applied to the phenomenon of progress towards the manifest goal of science. It involves inferring progress towards a latent goal because of the alleged explanatory power of that conclusion.

Popper has felt the pull of this argument himself remarking in 1974 that:

> it would be a highly improbable coincidence if a theory like Einstein's could correctly predict very precise measurements not predicted by its predecessors unless there is ... a higher degree of verisimilitude *than those of its competitors* which led to predictions that were less successful. (*RC*, pp. 1192–3)

Popper noted that there may be a **whiff** of inductivism in this argument. It is more than a whiff, this is a full-blown storm of induction. Popper thus faces a devastating dilemma. Either we accept this inductive argument to vindicate the claim of progress or we have to regard science as not being a paradigm of rationality. Clearly no argument other than an inductive one will do the trick. For there is no deductive argument from what we observe about science— about its predictive and manipulative success—to its verisimilitude where increasing verisimilitude means increasing truth at the theoretical level. But if we admit the legitimacy of induction here, why not admit it elsewhere? Having accepted any inductive argument as legitimate we have the falsification of the grand Popperian experiment.

But without accepting some such argument we have no reason to accept science as successful where success is defined in terms of verisimilitude. Indeed, we have no reason even to accept that science is probably successful. In which case we have no answer to our question because we are operating in a framework in which we cannot even recognize that science is successful.

One might be tempted to think that Popper would be better off sticking to what I called the manifest goal of science. Why not understand the goal of science simply in terms of increasing predictive power? There are serious arguments to support this move. But making this move is of no avail to Popper. For he cannot even recognize that Einstein is more successful than Newton in this sense. Let us suppose that Einstein has in fact passed many more attempted falsifications than Newton and that Einstein has failed no tests to date. To assume that Einstein is predictively better on that basis is to reason inductively. It is to assume that the set of tests is a fair sample of the totality of tests conducted by some omniscient being. But this is an inductive argument. For it is logically possible that, notwithstanding these observations, Newton is better. We are using our local observations as a reliable sample for the entire universe. Perhaps there are vast regions in which Newton fares better than Einstein.

Popper cannot provide an answer to our question, for he cannot even recognize that which puzzles us. Perhaps the answer lies in taking a yet more modest view of the goal of science. It is not rational to pursue a goal if you have good reason to think that you cannot even progress towards it. In such a situation the most prudent and most rational approach may be to change one's goal. Why not consider the goal of science to be the pursuit of unfalsified theories of high content? Well this is just too easy. We could invent endless theories of high content that are not falsified simply because no one bothers to try. Perhaps the aim is unfalsified theories which have survived serious attempts to falsify them? But why should we value such theories? Why should we aim at this? The only reason could be that we take their success in resisting falsification as evidence that they are likely to have more empirical success than their rivals. But here again we need the inductive step in order to make the framework acceptable.

We are forced to conclude that relative to the goal Popper specifies for science, science is not a rational activity given his non-inductivist methodology. I hasten to add that similar objections can be made with some force against other accounts of the goal and methods of science.

I have considered an internal difficulty in Popper. This is one of a number of deficiencies which vitiate the Great Experiment; that is, the construction of a theory of science which is thoroughly non-inductivist. It was however an important experiment in the theory of science. Some of the greatest achievements in philosophy are failed experiments. One form of such experiments is to take some concept very dear to us and construct a theory without using it. Popper drops induction, Quine drops meaning. The result does not convince us in the end. In the case of Popper and in the case of Quine it shows us more clearly than we were aware just why we need the concept in question.

At this juncture I want to highlight an evolution in Popper's thinking to which various writers have drawn attention. Popper shifted from focusing on the demarcation between science and non-science by reference to empirical falsification to the demarcating 'criticizable from non-criticizable theories' ('*PDR*', p. 95). He also writes:

> Empirical refutation, is just part of the general critical work of improving theories (including metaphysical theories) by searching for errors ('*PDR*', p. 98)

This shift goes hand in hand with a conception of rationality which is broader than that of scientific rationality. In *The Open Society and its Enemies* (p. 225) he wrote:

> rationalism is an attitude of readiness to listen to critical arguments and to learn from experience. It is fundamentally an attitude of admitting that '*I may be wrong and you may be right, and by an effort, we may get nearer to the truth*'.

And in *Unended Quest* (p. 117) Popper writes that he stressed:

> that the critical method, though it will use tests wherever possible, and preferably practical tests, can be generalized into what I described as the critical or rational attitude. I argued that one of the best senses of 'reason' and 'reasonableness' was openness to criticism—readiness to be criticized, and eagerness to criticize oneself; and I tried to argue that this critical attitude of reasonableness should be extended as far as possible.

This is a move in the right direction. To hold falsifiable beliefs and to hold them non-dogmatically is a general epistemic virtue. It is not particularly a scientific virtue.

One's beliefs about the character of one's friends should be falsifiable and non-dogmatically held. Someone who dogmatically holds that someone else is a coward, a saint or a crook is not reaching the ideal standards of epistemic rationality. So too should our beliefs about political and social processes be non-dogmatically held. To the extent to which this is so, we lose the possibility of explaining why science is so specially successful by reference to falsifiability as the key to the scientific method—it is a key to something much more general; good epistemic practice.

Empirical falsifiability and the absence of dogmatism, then, is a good thing in general but it is not exclusively a mark of scientific rationality. Popper in the passages quoted above and elsewhere goes further. His ultimate concern is with a broader kind of rationality. The decisive point of rationality, according to Popper, is the critical attitude, the attitude that I may be wrong and you may be right, and that together, we may, with effort, get nearer to the truth.

Popper is entirely right to move to a general notion of rationality and not to focus on a specific notion of scientific rationality. In so far as falsifiability might be what results when one proceeds rationally in science, my claim, with which he may disagree, is that falsifiability is a virtue but it is not a specifically scientific virtue.

If rationality is understood in this way then science is rational. For science is an institution in which criticism is the norm. But this is true even of inductivist scientists. For no one disputes the importance of criticism in science. The real dispute concerns the notion of positive evidence for the truth or likely increase in approximate truth of theories. The inductivist sees criticism as a tool in achieving his inductivist ends. For the Popperian it can appear either as an end in itself or as something which we take on simple blind faith as helpful in achieving increasing verisimilitude. The problem for the falsificationist is to explain why we should care about rationality if it is defined in this way—in terms of criticism. What value is criticism? Well it weeds out falsehood. But falsehoods are without number. Only if by weeding out falsehood we raise the chance of truth or verisimilitude can we have a reason for valuing criticism. The reconstrual of rationality in terms of criticism makes science—both inductivist and non-inductivist science—rational, but only the inductivist can explain why we should care about rationality thus defined.

This point generalizes. Accounts of scientific rationality typically fail to provide an account of something that is specifically scientific. I offer two illustrations. Kuhn in his restrained moments characterizes scientific method in terms of adherence to what he calls the scientific values. These are, Kuhn claims, constitutive of what it is to be a scientist: accuracy, consistency, scope, simplicity and fertility. I do not doubt that these are scientific values. But they are so in virtue of being general epistemic values. Few epistemic agents turn their faces knowingly against consistency. All things being equal we are attracted to moral theories which are simple. As an amateur chef I thought 'nouvelle cuisine' was fruitful. It was an approach which could be elaborated and applied to new phenomena—old dishes given a novel treatment. Of the theories which explain what animates students today I am attracted to one which is broad in scope:

namely, worries about jobs after graduation. That explains the higher work level and the popularity of law. And accuracy is as much a virtue in a furniture maker as in a scientist. Another illustration is to be found in the fashionable attempt to characterize the method of modern science by reference to inference to the best explanation. I gave examples of this strategy in the cases of Galileo and Einstein. Some philosophers of science have argued that progress in science has gone hand-in-hand with the methodological discovery that inference to the best explanation is a good thing. This idea is attractive. No doubt science has progressed because we are willing to posit the unobserved and even the unobservable in order to explain what we observe. But this is not in any way uniquely scientific. It is just part of good epistemic practice. We all use inference to the best explanation all the time. I believe I have moles in my fields in Wales because that explains the otherwise mysterious mounds of earth which keep annoyingly appearing. It is through inference to the best explanation that I have come to believe that the editor of this book is thoughtful and generous—how else to explain his offer of a nice dinner after the lecture on which this paper is based? And I believe that Robert Maxwell was devious and unscrupulous. How else to explain his apparent success for so many years? I did not observe these psychological states of our editor and of Robert Maxwell, I inferred them, inductively, in order to explain things I do observe.

It is my suspicion that any characterization of scientific method or scientific rationality given at the level of abstraction that philosophers offer will fit science as well as non-science. That being so, the attempt to explain the success of science by reference to the discovery of scientific method is going to be deficient.

I have elsewhere referred to those philosophers of science who offer these highly abstract and general characterizations of scientific methodology as the **Great Methodologists**. As I have indicated I have doubts about how much they contribute to the question that concerns me here. There is, however, a different sort of investigation which has more content and may be part of the answer we seek. This is the work of what I have called the **little methodologist** who offers specific characterization of work in specific areas of science. Such methodologists investigate procedures specific to particular areas of science, procedures which do not apply to all of science or indeed to all the activities within a particular science.

If I am right in this conjecture—that philosophers' characterizations of scientific rationality (including that of Popper), in so far as they are parts of the story, are parts of the general story of good epistemic practice. As such they cannot be the crucial ingredient in any account of the success of science in achieving its manifest aim.

What are we to do? The answer lies in Popper. In view of the importance of this insight I quote at length from his discussion of the sociology of knowledge in Chapter 23 of *The Open Society*:

> And, ironically enough, objectivity is closely bound up with the *social aspect of scientific method*, with the fact that science and scientific objectivity do not (and cannot) result from the attempts of an individual scientist to be

'objective', but from the *friendly-hostile co-operation of many scientists*. Scientific objectivity can be described as the inter-subjectivity of scientific method. But this social aspect of science is almost entirely neglected by those who call themselves sociologists of knowledge.

This passage is followed by a discussion of the possibility of some Robinson Crusoe succeeding as a scientist. Popper allows that while a Robinson Crusoe might do significant things that looked somewhat like science, it would not be real science any more than would the deliverances of clair-voyants who produced what looked like good scientific papers:

there is an element of scientific method missing, and consequently, that the fact that Crusoe arrived at our results is nearly as accidental and miraculous as it was in the case of the clairvoyant. ... What we call 'scientific objectivity' is not a product of the individual scientist's impartiality, but a product of the social or public character of scientific method; and the individual scientist's impartiality is, so far as it exists, not the source but rather the result of this socially or institutionally organized objectivity of science. (*OS*, II, pp. 219–220)

It is no longer true to say that this aspect of scientific method is neglected by sociologists. It is not so much that they neglect it, but that they misunderstand it (as I will indicate below). But it is true to say that it is neglected by philoso-phers of science.

Before developing this sketch, let us rehearse what we are seeking. Science is said to be successful. Sir Peter Medawar in a book with a trick title, *The Limits of Science*, writes that 'Science is incomparably the most successful enterprise human beings have ever engaged upon.' And Laudan tells us that science unlike other areas of human endeavour is excelling itself: its success is accelerating. These are deeply problematic claims. But for the sake of argument I leave their evaluation aside. What I will carry forward is the implicit invitation to compare science with other social activities.

What I am advocating might be called a **social epistemic** study of science in which we seek to identify aspects of the social practice of science which contribute to its achievement of its epistemic ends. As before, I will take focus on the manifest aim of increasing predictive success. The question is then whether we can identify features of the practice which contribute to the achievements of science. In particular I look to norms enforced by the community which can justifiably be claimed to be epistemically successful. What I have to say strikes me as simple-minded and obvious. The only reason for saying it is that this aspect of scientific practices has been neglected by those philosophers who seek to account for the success of science with the exception of Popper. Most of the features to which I draw attention may be subsumed under his description of friendly-hostile co-operation.

The first norm is the prohibition against suppressing data from one's co-workers. To sin against this is really bad. One is just not supposed to tear out those pages of one's laboratory notebook which go against the hypothesis that one has advanced in print. Clearly this norm does serve the epistemic ends of science.

In the case of each of my social norms, where I can, I will cite a contrasting social institution. For, as noted above, talk of the progressive character of science invites a comparison with other forms of social activity. In the case of this first norm, contrast institutions include those of diplomacy and politics. In the case of those institutions the suppression of data is a positive virtue. If I am sent to negotiate a peace settlement with Milosevic, it will not do to begin by outlining the long history of Serbian atrocities. Suppression of data is not forbidden, it is encouraged. When Sir Robert Armstrong was caught doing just that in the Spy Catcher court case he replied that he had merely been economical with the truth. His career did not suffer.

A related social norm of epistemic value in science is the prohibition against the invention of data. This too is really bad. You can get expelled from the club for that kind of thing. Indeed, it is such a sin that you can be posthumously expelled as Cyril Burt was from the British Psychological Association on suspicion of having committed it. Again it is clear that the public dissemination of invented data would inhibit the realization of the epistemic goals of science. But what is particularly important from the point of view of explaining the success of science is the system which enforces this norm. I do not have a contrast institution in this case because I have not been able to think of an institution in which the invention of data is considered a merit.

The more sophisticated of the Great Methodologists think, contrary to Popper, that methodology has a history and that methodological discoveries are part of the explanation of accelerating success. So too can we expect to find progress in regard to the development of the enforced norms. The third norm to which I would draw attention has a history. This is the requirement that scientific ideas be made public. It seems obvious that this serves progress. By being made public they can be critically discussed and depending on the up-shot serve to encourage or discourage further work in a cost-effective manner. In a word, scientific ideas have originators but not owners. The contrast institution is technology. In this case, if ideas were made public instead of being owned and protected by patents, the capital investments necessary to exploit the ideas would not be made.

In the early days of modern science this was not a norm. Mathematicians often sent colleagues theorems which they had proved but declined to provide the proof. So hours were wasted seeking proofs. In the early days of the Royal Society, pressure had to be put on members to present their ideas. Notoriously Newton never made public his alchemical work. We have come to appreciate the epistemic importance of putting ideas to the public gaze. And we have a host of devices which encourage this. You cannot get promotion for discoveries, you get promotion for published discoveries.

Not all of the social factors which serve the epistemic goals of science are socially enforced norms. Others concern the social structure of science. Science is relatively non-hierarchical. Individuals can pursue success and have it recognized relatively independently of their position in the scheme of things. Contrast this with the Chinese Communist Party where age is a prerequisite for recognition.

This feature has to be balanced against another and particularly important one; namely, the system for the recognition of experts. In so far as experts can be recognized, according importance to their judgments serves the epistemic ends of science. It is, for instance, argued that potentially wasteful work on cold fusion was terminated because of the interaction of the network of experts who disseminated their negative judgments throughout the community. This is not to say that science is a closed community in which criticism of those holding certain positions is utterly *verboten*. However, it is to acknowledge that science works best through a system of accreditation of experts whose judgment is to be accorded serious weight.

There is another feature which interestingly provides at the moment a counter-example to that favoured methodological rule: offer only falsifiable theories. This is the way in which the institution encourages innovation and deviance. To guard against the possibility that the orthodoxy is headed up a blind alley, a certain number of members of the institution are encouraged to swim against the stream. String theory is a case in point. It is not a falsifiable theory. It makes no predictions whatsoever. Recently an appointments committee for a post in theoretical physics was told by distinguished experts that the university in question should have a string theorist—just in case. I suppose the Catholic Church provides a contrastive institution in this regard.

For a final point on the social epistemics of science I turn to someone with whom I do not always agree, Richard

Rorty. Rorty has a well-known hostility to the Great Methodologists. But like most of us he is impressed by science and has sought to explain what is special about it by reference to the way scientists exhibit certain virtues; a commitment to using persuasion not force, incorruptibility and patience. I do not for a moment believe that scientists are conspicuously patient. Indeed, a certain impatience drives the epistemic engine of science. But I think he has a point about incorruptibility. You cannot easily buy them off. As he remarks:

> even today, more honest, reliable fair minded people get
> elected to the Royal Society than to, for example, the
> House of Commons. In America, the National Academy
> of Sciences is notably less corruptible than the House of
> Representatives.[4]

This is important. Rorty does not explore why it is that scientists have these virtues. And a serious study of this matter would have to consider the mechanism whereby the institution of science selects individuals with these characteristics. The social system of science discourages corruptibility and that is an important ingredient in its success.

These embryonic remarks do not explain the success of science. I seek to do no more than argue that if you seriously want to explain its success you must do more than deploy the rhetoric of Scientific Method and Scientific Rationality.

[4] R. Rorty, 'Is Natural Science a Natural Kind' in his *Objectivity, Relativism and Truth* (Cambridge University Press, 1991), p. 61.

Perhaps the writings of the Great Methodologists have a role to play in a negative sense. If their norms were not by and large respected science would go off the rails. But so too would any epistemic activity. And certainly what the Little Methodologist tells us is part of the explanatory story. But it is only part. The other part has come from what I have called social epistemics. This is a normative enterprise which seeks to identify aspects of the social practice which serves the epistemic ends of the enterprise. The further investigation of this requires a constructive partnership of philosophy and sociology.

I hasten to add that I am not in any way supporting the position of the radical sociologists of scientific knowledge such as the Edinburgh school. These radicals eschew any notion of the rational assessment of scientific theories. I am interested in a scientific investigation of those social aspects of science which contribute to its real, objective success. An example will serve to convey the contrast. For the radical sociologist of science, the notion of an expert is a social construct. For me the social process of identifying experts is a procedure for determining who really has expertise. For them there is no content to the notion of an expert beyond that of a socially recognized expert.

I have cast aspersions on the notion of scientific rationality as embodied in rational or philosophical models of science. I doubt that there is such a thing as scientific methodology as understood by such Great Methodologists. But I am a firm believer in the importance of rational assessments of our beliefs. We ought to strive to be rational in matters of belief. As scientists and as non-scientists we have good reason to apportion our degree of belief in a proposition to the strength of the evidence, and we ought to make appropriate efforts to gather and assess evidence (appropriate being a function of the cost of evidence and the importance of avoiding error). My doubts are about how accurate or explanatory it is to attempt to encapsulate what it is to be rational as a scientist in terms of the sort of characterization of scientific method offered by the Great Methodologists. It is time to dissolve that notion.

To re-capitulate my story here. Like Popper we are impressed with the success of science. It is tempting to think that that success might be explained by reference to a rational model. Popper's grand experiment of constructing a purely falsificationism model fails. What this amounts to is making the only constraint on scientific rationality deductive consistency. That neither fits with actual nor ideal scientific practice. Scientists like the rest of us are fundamentally inductive machines. The search for empirical falsification is part of the story, it is not the entire story. And in so far as it is part of the story it is not uniquely the scientific story. Popper's account like other accounts in so far as it captures a grain of truth captures a grain of truth about general epistemic activities. In moving to a plane of rationality in general Popper implicitly acknowledges this. Once this is acknowledged we have to look elsewhere for an explanation of what is special about science. Popper's remarks about the social dimension of rationality point us in the right direction. We can account for the specialness of science by reference to unique or relatively unique features of the social practice of science. We no longer need to represent the scientist as the Hero of Legend: heroics are not needed, the social structures do the work instead.

Before concluding I want to remark on the importance of Popper's generalized stand on rationality from more personal experience. It is perhaps too easy for us in relatively open western society to fail to see the significance of his stress on criticism. I give two illustrations. Some years ago, before the events of Tiananmen Square, I was involved in organising a meeting of philosophers in Wuhan. This became a politically sensitive event. Wuhan University withdrew sponsorship but the philosophy department did not. What emerged from our discussions was the political importance of Popper in China. Time and time again Chinese reformists argued that Popper's work needed to be made easily available. As they said, we do not have the tradition of criticism. We need that to bring reform; we need that to prevent the excesses of the Cultural Revolution ever occurring. Unfortunately the events of Tiananmen Square intervened making the need all the more clear.

Of course my own inductivist leanings mean that I am bound to say that criticism is only part of the story. Criticism is best if based on beliefs for which we have positive grounds. To elaborate I want you to imagine two social communities one populated by Popperians, the other populated by Humeans. Both communities attach with

Popper the highest importance to criticism. For the Popperian community there can be no requirement that the critic have grounds for his or her criticism. For the only grounds would be inductive ones.

The Humeans recognize that ultimately there are no grounds for anything. But except for philosophical periods on Sunday afternoons of philosophical reflection, they think and act as if induction was a rational procedure. Consequently, they proceed under the general constraint that the onus is on the critics to have grounds for their criticism and to make those grounds available. That is, the norm is to criticize by citing something for which you are prepared to give grounds. For me the Humean community is the more viable. We simply have not the time and energy to consider every objection. We proceed by attaching weight to those criticisms for which there are grounds. If I criticize the government on the grounds that less money is spent on higher education than in other European countries, my criticism deserves more attention if I have grounds for thinking (grounds which can only be inductive) that the government is in fact spending less.

It will no doubt be objected that I am not playing fairly. On the one hand, I criticize Popper for not being able, as an anti-inductivist, to provide any reason for thinking that there is progress in science. And I have also said that he can give no justification for encouraging criticism. For the only possible justification would have to be an inductive one. But on the other hand, I have not answered Hume's objection. I have not provided a vindication of inductive reasoning. The Popperians have faith that science progresses. I have faith in induction. Have we not reached that sort of stalemate where, to quote Wittgenstein, all we can do is to call the others 'dolts and heretics'?

Pending a convincing answer to Hume's scepticism which I fear is not likely to be forthcoming (at least from me), I suggest we conduct an experiment. Let us encourage at the epistemic level the policy of openness and pluralism that Popper has rightly argued for in the political and social spheres. We could create an epistemic community in which there were thorough-going falsificationists and committed inductivists. Or, perhaps these should be two communities to give a fairer test. It is my conjecture—a bold conjecture the testing of which I look forward to—that the inductivist community will have more to offer us. For that community will place the same stress on criticism but it will also have certain additional epistemic instruments to use.

Of course I cannot offer any grounds for my conjecture. What I have hoped to do is to offer an experiment; an experiment which can be recognized by both sides as legitimate.

For all my philosophical criticisms of Popper, no philosopher of the twentieth century has had or will have had a more beneficial cultural influence. And we still have to learn some of the lessons that Popper has taught. We do not subject our views, our assumptions, our beliefs to the kind of criticism he advises. In Sarajevo, students are shot at on the way to philosophy lectures. The Philosophy Faculty is located on the notorious 'snipers alley'. Many students have been killed. When you are there, in the philosophy building, with the bullets striking, you are asked why we permit this.

At that moment you become painfully aware that what we say to ourselves to explain our inaction does not stand up to criticism. And the truth is worse. We do not as a community subject the views of ourselves and our politicians to the criticism Popper called for. If we did we would not be where we are in Bosnia today.

2 Popper and Reliabilism

PETER LIPTON

K arl Popper attempted to give an account of scientific research as the rational pursuit of the truth about nature without any appeal to what he took to be the fictitious notion of non-demonstrative or inductive support. Deductive inference can be seen to be inference enough for science, he claimed, once we appreciate the power of data to refute theory. Many of the standard objections to Popper's account purport to show that his deductivism actually entails a radical scepticism about the possibility of scientific knowledge. Some of these objections appear unanswerable in the context of the traditional analysis of knowledge as justified true belief; but this is neither a conception of knowledge that Popper himself accepted nor one that is currently in fashion. Reliabilism, the view that knowledge is a true belief generated by a reliable method, is now a popular replacement for the traditional analysis and one that is closer to Popper's own conception of knowledge. My aim in this essay is to consider in brief compass the prospects of a reliabilist reading of Popper's account of science. Such a reading makes it possible to turn some of the standard objections and helps to show which of Popper's views should be accepted and which rejected.

The Standard Objections

Popper's philosophy of science is naturally seen as a radical response to Humean scepticism about induction. According to the sceptical argument, no form of non-demonstrative reasoning is rationally defensible, since any argument to show that such reasoning is generally truth-preserving or reliable would itself need to be a non-demonstrative argument and would hence beg the question. Our inductive practices have presumably been at least moderately reliable up to now, else we would not be here to consider the problem, but what needs showing is that they will continue to be reliable in future. The claim of future reliability is however a prediction that could only be justified inductively. What we have observed hitherto does not deductively entail that induction will work in future, but to give an inductive justification for that prediction is to argue in a circle. Our situation seems analogous to that of a party of hikers who could only get across a wide chasm with the aid of a rope neatly coiled on the far side.

Hume's own response to his sceptical argument is that inductive inferences are rationally indefensible but psychologically unavoidable. He goes on to give a description of what he took the psychological mechanism behind them to be, a process of Pavlovian conditioning or habit formation. Most epistemologists have instead rejected Hume's scepticism about induction and have attempted to show by more or less desperate means what is wrong with his remarkably resilient argument. Popper, by contrast, simply accepts the sceptical argument: induction is irrational. Unlike Hume, however, he does not retreat to the descriptive psychological project. Instead, Popper sets out to show that, scepticism about induction notwithstanding, scientific inquiry is rational, by showing that, appearances to the contrary, scientific inferences are purely deductive.

Popper's central idea is that although the scientific evidence never entails that a theory is true, it may entail that the theory is false. If we have a hypothesis of universal conditional form, no number of positive instances will entail that the hypothesis is true, but a single negative instance will entail that it is false. No number of black ravens entails that the hypothesis that all ravens are black is true, but a single white raven entails that the hypothesis is false. More generally, if a theory entails a prediction and the prediction is found to be false, then the theory must be false as well, since to say that the argument from theory to prediction is deductively valid is just to say that, if the conclusion is false, at least one of the premises must be false as well. Scientists can thus know that a theory is false, without recourse to induction. Moreover, faced with a choice between two competing theories, they can exercise a rational preference with respect to the goal of discovering the truth, if one of the theories has been refuted but the other not, since it is rational to prefer a theory that might be true over one known to be false. Induction never enters the picture, so Hume's argument is defused.

Popper thus seeks to accept Humean scepticism about induction without accepting scepticism about science. Many of the standard objections to Popper's position attempt to show that he cannot have it both ways: insofar as his account really does abjure induction, it makes scientific knowledge impossible. Let me remind you of four of the most familiar of these objections. Firstly, according to Popper scientists are never justified in believing that the observation statements they accept are true. Scientists adopt certain procedures for accepting data that they hope will lead to accepting mostly true statements, but a hope is not a reason. The only thing that could justify an observation statement would be the scientist's experience but, according to Popper, only statements can justify statements. Consequently, if an accepted observation statement contradicts a hypothesis, we are not justified in claiming that the hypothesis is false, only that either the hypothesis or the observation statement is false. On a traditional conception of knowledge, this makes it impossible to use the mechanism of empirical refutation to gain knowledge that any hypothesis is false. The data cannot be known to be correct and as a result theories cannot be known to be false. This is the problem of unjustifiable data.

The second standard objection is that knowledge of the falsity of a theory remains impossible even if we grant knowledge of the truth of the data. This is the problem of the auxiliaries. Theories do not entail predictions outright, but only with the aid of various often ill-defined auxiliary assumptions. From a logical point of view, what follows from the falsity of the entailed prediction is only that at least one of the premises is false, not which ones. Consequently, since the premise set includes statements apart from the theory under test, knowledge of the falsity of conclusion does not make it possible to know that the theory is false.

The third objection is the problem of application. Even if we granted that the mechanism of refutation made it possible to know that certain theories are false, Popper's account still does not make possible the rational application of science. In particular, we would have no reason to prefer for practical purposes the predictions of unrefuted theories over those of their refuted rivals. The source of the problem is the inapplicability of Popper's argument for rational preference between theories to preference between their predictions. According to Popper, we are to prefer the unrefuted theory, because it may be true whereas its refuted rival is known to be false. Every set of false statements has however indefinitely many true consequences, so we cannot say that the prediction from the false theory is itself false, and the basis for preference is lost. It is impossible to know that any of a theory's predictions are true, or that the prediction of one theory is more likely to be true than the incompatible prediction of another.

The final objection is that Popper's account provides no reason to believe that science is moving towards the truth. This is the problem of the bad lot. However assiduously scientists may eliminate and replace false theories, there is no reason to believe that the new theories are better than the old ones. Perhaps all the theories we will ever generate are false, and perhaps from among those it is the ones we fail to eliminate that are furthest from the truth. It is impossible to know that later theories are better than the ones they replace.

If these objections are sound, Popper's philosophy entails a profound scepticism about science. Nor is this something a Popperian ought to accept with equanimity: we must not confuse scepticism with fallibilism. A fallibilist account of knowledge is the ground we all wish to occupy: neither theory nor data are ever certain. Hume, however, was no mere fallibilist about induction. He did not claim that the conclusions of inductive inferences are uncertain: he claimed they were epistemically worthless. Similarly, if the objections to Popper's philosophy are sound, they show that the results of science are worthless, not merely uncertain.

Popper was well aware of the objections I have mentioned, but his replies are unsatisfying. His response to the problems of unknowable data and of the auxiliaries is, to put it baldly, that scientists should pretend they do not exist. Scientists should pretend that the observations statements they accept are known to be true, and they should pretend that a failed prediction refutes the theory directly. Even if this were good advice to scientists, which is doubtful, it does not meet the epistemic difficulties the objections raise. Popper's reply to the problem of practical preference is to claim that, if we must rely on any theory, it is rational by definition to rely on the best-tested one, which is one that has not been refuted. But if we are careful to free the notion of 'best-tested' from any inductive associations, this semantic solution is no more satisfying than the parallel semantic solution to the Humean problem, according to which using induction is part of what we mean by acting rationally. In both cases, the natural response is that, if this is what 'rational' means, what we care about is not being rational, but being right. Finally, Popper's reply to the problem of the bad lot seems to be that we may hope that his method of conjecture and refutation takes scientists towards the truth and that there is no better alternative. But scepticism is not avoided by calling it unavoidable, and it is unclear on what basis Popper could argue that his method is any more likely to generate true theories than random guessing.

Reliabilism

The standard objections to Popper's account of science have considerable force. Each of them casts serious doubt on the power of a Popperian methodology to generate scientific knowledge. In each case, the argument for the impossibility of knowledge is based on the impossibility of justification. This suggests that at least some of the objections might be turned by using an analysis of knowledge that does not depend on the notion of justification, a thought that fits well with Popper's general hostility to such a notion. In recent epistemology, the most discussed such analysis is reliabilism. Can Popper's philosophy of science be improved by combining it with a reliabilist theory of knowledge?

Reliabilism was not originally developed as a response to Humean scepticism, though it was later so applied. Instead, it was motivated by the thought that having a justification seems neither necessary not sufficient for a true belief to count as knowledge. Perceptual knowledge is the obvious example of a kind of knowledge that does not require justification, at least if justification is understood as explicit argument. Over-intellectualizing philosophers to the contrary, when I see that my pen has fallen on the floor, my knowledge of its present location is not based on argument or inference. Even cats and dogs have perceptual knowledge, though they are incapable of giving reasons. Other plausible examples of knowledge without justification include knowledge from memory and, at least for humans, from testimony. There can be knowledge without justified true belief.

Conversely, the standard source of examples for justified true belief without knowledge are the famous Gettier cases. Gettier constructed a conceptual machine for generating examples of justified true beliefs that are not knowledge. The machine works simply by deducing true beliefs from justified but false beliefs, capitalising on the logical point, already noted above, that every false statement has innumerable true consequences. These derived true beliefs will not in general be cases of knowledge. For example, suppose I had the justified but, as it happens, false belief that my wife took our car to London for the day, from which I deduced that it would not be at our house when I came home. The deduced belief was true as it happens, but only because our car was stolen during the day. My belief that the car would not be at home was true and justified, but not a case of knowledge.

According to reliabilism, the justification condition that a true belief is knowledge just in case one has good reasons for the belief should be replaced by the condition that the true belief was generated by a reliable method or process. Knowledge is reliably produced true belief. Different reliabilists have analysed the notion of a reliable method in different ways. For example, some have understood a method to be reliable just in case it tends to produce true beliefs, others have construed reliability in terms of

PETER LIPTON

various counterfactuals concerning the resultant belief, for example that one would not have held the belief, had it been false. Not all true beliefs are knowledge, since it may be a matter of luck that the belief is true, but if the truth was reliably produced it counts as knowledge, even if the reliable method did not involve argument or justification. Thus if our perceptual mechanisms or those of some other animals reliably generate true perceptual beliefs, those beliefs count as knowledge. And a justification may fail to generate knowledge by failing to satisfy the appropriate reliability condition. I did not know that our car would not be at the house, in spite of having a good reason to believe this, because it was a fluke that my belief was true, and this is reflected in the fact that I would still have had that belief if the car had not been stolen.

How best to articulate the notion of reliability suitable for the analysis of knowledge is an interesting and difficult question which, though clearly central to the development of a reliabilist theory of knowledge, is one I hope to avoid in this essay. There are, however, three general features of reliabilist accounts of knowledge that are worth emphasizing. The first is that reliabilism does not drop the truth condition on knowledge. To say that a method is reliable is to speak of its propensity to generate true beliefs, not merely beliefs that are useful or otherwise attractive, whatever the other connotations of the term 'reliable'. Secondly, what counts for knowledge is reliability in fact, not having reasons to believe that one's methods are reliable. The cat knows birds when it sees them, but of course can give no reason to believe that its visual system is a reliable bird-detector. Finally, although reason and inference are not required for knowledge, there is such a thing as inferential knowledge. What makes this knowledge, according to the reliabilist, is just that the methods of inference employed are in fact reliable. Inference is one set of methods for generating beliefs among others.

Although not constructed for this purpose, reliabilism offers a quick non-Popperian solution to scepticism about inductive knowledge. Suppose that Hume is right in claiming that there is no possible justification for the claim that our inductive practices are reliable. Even so, from a reliabilist point of view, the impossibility of knowledge does not follow. For inductive knowledge to be possible, it must be possible that our inductive practices are reliable in fact, a possibility Hume does not deny. Certainly he could not argue that our inductive practices will be unreliable, since this would be a prediction that could itself only be justified inductively. So inductive knowledge is possible and if, as we all believe, induction is at least a moderately reliable method of acquiring beliefs, then it is actual as well.

The Possibility of Negative Knowledge

My topic in this essay, however, is not how reliabilism may vindicate induction, but rather how it may vindicate Popper's emphasis on negative methods, by helping to turn the four standard objections to his position. Let us begin with the first two, the problems of unjustifiable data and of the auxiliaries, both of which threaten the claim that scientists can know that the theories they reject are false. Popper would have liked falsification to have the certainty of proof, as his use of the word 'refutation' suggests. Such certainty is possible in the relation of logical incompatibility and Popper would have scientists act as though the rejection of theories has a similar status when motivated by that relation. But on Popper's view scientists do not know that the data they accept are true or that the theory is to blame for failed predictions, so they cannot know that the theory they reject is false. From a reliabilist point of view, however, falsification becomes reliable rejection. If using the deductive relation enables scientists reliably to reject falsehoods and not to reject truths, then they can know that the theories they reject are false. Neither the unjustifiability of the data nor the presence of additional premises rules this out.

The unjustifiability of data by experience poses no special difficulty for the reliabilist, since knowledge does not require justification. At the same time, the reliabilist allows that experience is more than a motivation for accepting observation statements, since it is part of the method which causes those statements to be accepted. On this view, the senses are detection devices which generate knowledge insofar as they are reliable. This is not to show that they are reliable, but such a demonstration is not required in order for the data to be known. Reliabilism does not show that scientists do know that their data are correct, but it improves the basic Popperian scheme by showing how such knowledge is possible and by explaining the epistemic relevance of experience, without abandoning the Popperian stricture against the notion of the justification of statement by experience.

Reliabilism also helps with another challenge to the possibility of knowing the data that Popper was among the first to identity, the theory-ladeness of observation. For the reliabilist, the essential role of theoretical beliefs in the generation of data is no bar to knowledge of the data. Of course if the lading theories are false and are so related to the data that this entails that the data must be false as well, then the data cannot be known. But the presence of theories in the generating mechanism is compatible with the reliability of that mechanism. For example, when scientists use theories of their instrumentation to get data from those instruments, what counts as far as knowledge is concerned is whether the composite mechanism, consisting of both physical and intellectual technology, tends to generate only correct data. The reliability of this mechanism does not depend on a theory-neutral description of the evidence.

The reliabilist takes a similar line on the problem of auxiliaries. Where premises apart from the theory under test are needed to deduce a prediction, the falsity of the prediction does not entail the falsity of the theory, but this is no bar to using the Popperian method of refutation to gain knowledge of the falsity of theory. What counts is that the scientists' attribution of blame be reliable: that they tend to blame the theory only when it is to blame. Popper has often been criticized for saying that scientists ought generally to blame the theory rather than the auxiliaries, but this is a criticism the reliabilist can easily absorb. Scientists have complex and poorly understood ways of apportioning blame, of deciding whether to apply it to the theory or to auxiliaries and to which part of either, but insofar as these practices are reliable, they can yield knowledge of what is false.

The problems of unjustifiable data and of the auxiliaries show that falsification, unlike incompatibility, cannot be understood as a purely deductive relation. A reliabilist account of knowledge shows how one may nonetheless claim that scientists can know that the theories they reject are false. We should now ask to what extent this reliabilist picture of the rejection of theories remains recognizably Popperian. It does appear to be able to respect Popper's proscription on justification. Knowledge of the falsity of a theory does not depend on the justification either of observation statements or premises not rejected. But can the reliabilist respect Popper's asymmetry between confirmation and refutation? Reliabilism is in itself not hostile to induction and, as the problems of the data and the auxiliaries show, neither the reliabilist nor Popper can make sense of refutation as a purely deductive operation. Nevertheless, several important asymmetries between positive and negative methods remain from a reliabilist point of view. To put the matter crudely, the reliabilist can explain why falsification is easier than verification, why it is easier for scientists to acquire reliable methods for determining that a particular theory is false than for determining that a theory is true.

Popper's logical asymmetry is based on a contrast between part and whole. To determine that a universal generalization is true, we need to know about all its instances, whereas to determine that it is false we need only know about one. Similarly, to determine that a theory is true we need to know the truth value of all its consequences, whereas we need only know that one consequence is false to determine that the theory is false. This asymmetry survives the complications of falsification that we have considered and the reliabilist response to them. Whatever the method for determining the truth value of a statement, it will be easier to determine the truth value of one than of many, at least where the many includes the one. Similarly, it is easier to construct a reliable method for detecting the truth-value of some of the consequences of a theory than of all of them.

There is another source of asymmetry between negative and positive methods on the data side that the reliabilist can account for. Judgements of inductive support must satisfy the 'total evidence condition': they must be made with respect to all the available evidence. This is so because, unlike deductive reasoning, inductive reasoning is 'non-monotonic'. If a deductive argument is valid, it will remain so whatever additional premises are added. By contrast, an argument that we judge to be inductively strong may no longer be so judged when additional data are added. To take a Popperian example, having seen many black ravens I may, if I indulge in induction, infer that all ravens are black, but I will retract the inference if I later see a non-black raven, without rejecting any of my earlier data. This is why inductive assessment must be made in light of all the available evidence. The situation is different with respect to falsification. Having convinced myself that I have found a genuine counterexample to my hypothesis, I do not need to take into account all the other available evidence: more white ravens will not undo the damage caused by the black ones. From a reliabilist perspective, this provides another reason why we should expect it to be easier for scientists to construct reliable methods of rejection than of acceptance, since it is easier to construct a method that can use limited inputs than one that must accept and assess all the available data in one go. Negative methods can thus be seen to have a double advantage over positive ones, requiring less output and less input to be effective.

Falsification is Necessary for Positive Knowledge

I have argued that although Popper's account of falsification makes knowledge that a theory is false impossible on a justificationalist account of knowledge, a reliabilist account makes such knowledge possible and also helps to explain why it should be easier to falsify than to confirm. What I want now to show is that falsification is not just possible but also necessary if there is to be any positive scientific knowledge, since any reliable method of scientific discovery depends on the reliable elimination of false hypotheses.

Given that reliabilism in no way excludes inductive inferences, one might wonder why there couldn't be a method yielding positive scientific knowledge that did not depend essentially on the elimination of falsehood. The history of science provides overwhelming evidence that science does in fact depend on elimination, but there are also reasons of principle why this should be so. One is a central Popperian theme, the impossibility of inductivism. Most scientific theories appeal to entities and processes not mentioned in the evidence: those theories are not simple extrapolations and interpolations. Hence there is no algorithmic route from data to theory. Moreover, again as Popper has emphasized, a theory usually needs to be generated before collecting the data that best tests it, since the scientist can often only tell which data are relevant in light of that theory.

There is a world of difference between rejecting inductivism—the view that there is a mechanical procedure for moving from available data to theory best supported by that data—and rejecting the possibility of induction or non-demonstrative support. Popperians have not always been careful to distinguish the two. Nevertheless, the impossibility of inductivism and the temporal priority of theory to data shows the necessity of negative methods.

If there is no mechanical route from data to theory and theories must be generated before scientists can find the data that would test them, there is no non-miraculous way for scientists to generate only true theories. Hence scientists must rely on methods of elimination, however much they may also depend on inductive support. Moreover, since the method of generation is bound to throw up some false theories, it must be designed so that the false theories that are generated can be reliably eliminated, if positive knowledge is to be possible. This is the reliabilist's version of Popper's falsifiability requirement on scientific theories.

Negative methods are thus an essential feature of any way of doing science that would yield positive knowledge. There may even be a greater role for negative methods in science than Popper's own account suggests, because of the constant use of negative filters in hypothesis generation.

This is suppressed in Popper's discussion, because of his artificial separation of the contexts of discovery and evaluation and his consequent neglect of the process of theory-generation. There are strong constraints on generation, including just the same mechanisms that later lead to the elimination of some theories that are generated. Many theories never make it to the testing-stage, because the scientist who thinks of them can eliminate them on already available information. Theories that nobody generates can in one obvious sense not be eliminated, but many of the constraints on generation are equivalent to those of elimination.

Falsification is Sufficient for Positive Knowledge

Reliable falsification is thus necessary for positive scientific knowledge; I now want to suggest why, perhaps more surprisingly, it is also sufficient. Let us suppose that scientists' eliminative methods are in fact reliable, yielding knowledge that various theories are false. What follows from this? Recall the problem of the bad lot: if all the theories we generate are false, then eliminating some of them won't leave us with the truth. Could not scientists be perfectly reliable falsifiers, never rejecting what is true, yet be quite hopeless at generating truths? They would then have plenty of negative but no positive knowledge, which would still be a form of scepticism about science. At first glance, this possibility appears to remain open, but this is an illusion: if scientists are reliable falsifiers, they must also be getting at the truth.

The connection between negative and positive knowledge is hidden on a naive deductivist picture. If scientists had some theory-independent way of determining the truth of the data, and some of those data contradicted the theory outright, then they could reliably eliminate some false theories even if they had no way of determining which theories are true or indeed even if none of the theories scientists generate are true. Popper never thought this was our situation, but his tendency to have us pretend as if it were has suppressed the connection between negative and positive knowledge.

For falsification to be reliable, scientists must know that the data are true: the method of data acceptance must itself be reliable. But given the extent to which the method of accepting data depends upon background theories, and given the extent to which the data are, as Popper emphasized, theory-laden, reliable data acceptance is only possible if our background is largely correct.

The same conclusion follows from the role of auxiliary statements in testing. Unless those premises were largely true, a practice of laying blame on the theory under test would not be reliable. In practice, the situation is more complex since, as we have already noted, scientists often do not blame the theory for a failed prediction. This however strengthens the present point. Scientists' selective attribution of blame could not be reliable unless they were reliable judges of when the background is correct and when it isn't. To put the point in general terms, elimination requires not just that scientists accept data but also that they accept theory, and the elimination can only be reliable if the acceptance is.

This argument does not show that science does take us to the truth: it does not solve Hume's problem. What it does purport to show, however, is that knowledge of which theories are true and knowledge of which theories are false are intimately connected. Popper may have been over-optimistic about scientists' ability to detect falsehood, but he was also over-pessimistic about what follows from this about the ability to detect the truth. Scientists cannot be good at detecting one without being good at detecting the other.

What then of the asymmetry between negative and positive methods? If what I have just claimed is right, that scientists could not know what is false unless they also knew what is true, then how could what I have said earlier be right, that it is easier to determine that a theory is false than that it is true? The answer lies in a difference of scope or scale. It is in general easier for scientists to show that a particular hypothesis is false than that a particular hypothesis is true, because showing falsehood requires the determination of the truth value of only a part of the hypothesis and because it requires only the use of part of the available evidence. Nevertheless, a method of falsification can be generally reliable only if there are also in use reliable methods of generating true beliefs concerning both data and theory.

Whence Induction?

I have suggested three main ways that a reliabilist account of knowledge may help to vindicate the Popperian emphasis on a negative methodology of science, in the face of various standard objections to Popper's position. Reliabilism shows how falsification is possible, why it is necessary for positive scientific knowledge and why it is sufficient. It cannot however be used to defend Popper's wholesale rejection of induction.

Induction is just non-demonstrative inference. It cuts across the distinction between acceptance and rejection, between positive and negative methods. Some acceptance, such as acceptance of observation statements, could in principle be non-inductive because, though inferential, the falsity of the claim is strictly entailed by the truth of the data. In real cases, however, the acceptance or rejection of a scientific theory involves non-demonstrate inference. The scientist has reasons for the judgment, and those reasons are inconclusive.

There is no reliable route to falsification that does not use induction. As we have seen, scientists also need inductive methods that yield positive results if their negative methods are to be reliable. Moreover, there is in my view no adequate response to the problem of application that does not concede positive inductive argument. To deny that scientists ever know that any of the unobserved consequences of their theories is scepticism. To accept that such knowledge is possible is to accept inductive inference to a prediction, however heavily mediated it may be by the technique of conjecture and refutation.

Induction is unavoidable, so Popper's solution to the problem of induction fails. But Popper's case against inductivism stands, as does his emphasis on the importance of a negative methodology. Reliabilism shares Popper's focus on the search for truth, takes the unavoidability of induction in its stride, and explains why Popper was right to put such weight on the role of theory elimination in science. It also brings out the way in which negative and positive results come together in science. If science generates knowledge at all, it can only do so by determining what is false; but if it can determine what is false, it can also determine what is true.

3   The Problem of the Empirical Basis

E. G. ZAHAR

1.   Popper's Demarcation Criterion

In this paper I shall venture into an area with which I am not very familiar and in which I feel far from confident; namely into phenomenology. My main motive is not to get away from standard, boring, methodological questions like those of induction and demarcation; but the conviction that a phenomenological account of the empirical basis forms a necessary complement to Popper's falsificationism. According to the latter, a scientific theory is a synthetic and universal, hence unverifiable proposition. In fact, in order to be technologically useful, a scientific hypothesis must refer to future states-of-affairs; it *ought* therefore to remain unverified. But in order to be empirical, a theory must bear some kind of relation to factual statements. According to Popper, such a relation can only be one of potential conflict. Thus a theory T will be termed scientific if and only if T is logically incompatible with a so-called basic statement b, where b is both empirically verifiable and empirically falsifiable. (We shall see that neither the verifiability nor the falsifiability of b was meant, by Popper, in any literal sense.) In other words: T is scientific if it entails ⌐b; where b, hence also ⌐b, is an empirically decidable proposition.[1]

This demarcation criterion which provides the best explication of the scientific character of theories, not to say an essential definition of empirical science, has been subjected to two very different types of criticism; the first is levelled at its theoretical and the second at its empirical aspect. Let us start with the first kind of criticism which is often referred to as the Duhem–Quine problem.

2. The Duhem–Quine Problem

Duhem,[2] then Quine,[3] rightly pointed out that no fundamental hypothesis is ever tested in isolation; it is always taken in conjunction with auxiliary assumptions and with descriptions of initial conditions. Lakatos added that one of these assumptions is a so-called 'ceteris paribus clause', which consists not of a single sentence but of an infinite or indefinite number of caveats. Should we face a refutation of some scientific system, we would thus not know what had actually been falsified; we should certainly have no reason to suppose that the fault lies with our core hypothesis, i.e. with the proposition we set out to test.

This objection to Popper's demarcation criterion, though partly justified, does not appear to me to be very

---

[1] See Popper: *BG*, p. 125.

[2] P. Duhem, *The Aim and Structure of Physical Theory*, Part 2, Chapter 4 (Princeton, 1914).

[3] W. v. O. Quine, *From a Logical Point of View*, Chapter 2 (Harvard, 1961).

damaging. If we subscribe both to hypothetico-deductivism and to first-order logic—as Duhem, Quine and Lakatos effectively did—then by the compactness theorem: if ⌐b is a logical consequence of any set S of premises, ⌐b already logically follows from a finite subset F of S. Taking T' to be the conjunction of all the elements of F, we can affirm that T', which may well be stronger than our core hypothesis T, can be falsified by b. That is a refutation always hits a finite proposition of which T may form part.

Let us now turn to the second type of criticism, which was levelled at the empirical aspect of Popper's demarcation criterion.

3. The Problem of the Empirical Basis

The refutation of T' would thus be definitive if b could be verified. But this is the point at which the second type of criticism of Popper's criterion sets in; a criticism largely based on Popper's own *conventionalist view* of the empirical basis.

Following John Watkins,[4] I shall regard a singular sentence p as expressing a level-0 proposition if p describes, in the first person, the immediate contents of a speaker's consciousness. Examples of such sentences are: 'I feel pain (now)', 'I seem to be perceiving a red patch', 'It seems to me that the arm of the galvanometer (that is what I take to be the arm of the galvanometer) has just moved', 'I seem to be perceiving a group of pink elephants', etc.

[4] J. W. N. Watkins, *Science and Scepticism*, Chapter 3. (Princeton 1984).

I shall take it for granted that *the truth-value of such a proposition p is logically independent of all transcendent states-of-affairs,* i.e. of all events occurring outside the speaker's consciousness. This is why level-0 statements are also referred to as immanent, autopsychological or phenomenologically reduced sentences.

Before going into the details of Popper's position, let me define it in broad terms. It consists of the thesis that the basic statements which test scientific theories are not and should not be of the autopsychological kind; they should rather consist of objective, i.e. of intersubjectively ascertainable propositions. Because of their reference to external objects accessible by means of theory alone, such basic statements are not verifiable; they are temporarily accepted on the basis of conventions regulated by well-defined procedures. Such conventions have in themselves nothing to do with the truth-value of the accepted basic statements.[5] This is why Popper's view of basic statements will henceforth be referred to as the *conventionalist thesis.*

Both John Watkins[6] and myself[7] have independently arrived at the conclusion that Popper's conventionalist thesis is mistaken; also that level-0 statements play an important role as explananda of certain theories. I say 'independently' not in order to avoid an anyway unlikely priority dispute, but to indicate that our positions were largely

[5] See Popper: *BG*, p. 125.

[6] Watkins, *Science and Scepticism*, Chapter 3.

[7] E. G. Zahar, 'The Popper–Lakatos Controversy', Section 2, in *Fundamenta Scientiae*, Vol. 3, No. 1, pp. 21–54. (Pergamon, 1982).

dictated, or rather constrained, by the intrinsic problems besetting Popperian methodology. John Watkins and I still differ over the nature of the empirical basis, especially that of the strictly physical as distinct from the psychological or psycho-physical sciences. John Watkins holds that level-1 statements form the basis of physics while level-0 propositions are the exclusive concern of psychology and psychobiology. Thus, according to Watkins, autopsychological statements lie outside the proper domain of physics; but they can be appealed to, as a last resort, in cases of litigation.

I have defended a much riskier view; namely that only autopsychological reports should count as scientific *basic* statements. Let me however start by playing Popper's advocate; i.e. let me show that Popper had solid reasons for rejecting level-0 sentences as suitable candidates for membership of the empirical basis. Popper argued as follows.

[A] Level-0 statements are incorrigible or indubitable only in the sense of not being psychologically doubted. But we know that feelings of conviction, no matter how strong, are often misleading: they neither establish nor even probabilify the propositions at which they are directed.

> I admit again that the decision to accept a basic statement and to be satisfied with it, is causally connected with our experiences—especially with our *perpetual experiences*. But we do not attempt to *justify* basic statements by these experiences. Experiences *can motivate a decision*, and hence an acceptance or rejection of a statement, but a basic

statement cannot be *justified* by them—no more than by thumping the table.[8]

[B] Autopsychological reports have the added disadvantage of being private to the person uttering them. Their veracity cannot therefore be checked, for they are not subject to repeated testing; hence no intersubjective agreement regarding either their truth-value or even their general acceptability can be reached.

[C] Psychology is an empirical science which investigates— among other things—the occurrence of feelings of certainty with regard to some statements; it demarcates between the cases where such feelings mislead us and those where they are more or less justified. Thus, testing psychological hypotheses by means of level-0 statements supported exclusively by feelings of conviction would land us into a vicious circle. We would naturally trust only those experiences of certainty which, according to the hypothesis under test, are justified. But the hypothesis might be questionable, which is precisely why we may have decided to perform the test in the first place.[9]

[D] It is therefore preferable to choose, as elements of our empirical basis, intersubjectively testable singular sentences. Such sentences will consequently refer to publicly observable entities, more particularly to physical objects which can be inspected by different experimenters and

---

[8] Popper: *LSD*, p. 105. For the rejection of the view that experiences confer a degree of certainty onto protocol statements, see *LSD*, p. 104, footnote 1.

[9] See Popper, *BG*, p. 125.

at different times. Such objects are transcendent in the sense of lying beyond the reach of the observer's consciousness; so they are grasped largely by means of theory. This is why basic statements are fallible on at least two counts. On the one hand, they refer to external objects which may either be non-existent or else possess properties different from those attributed to them by the observer; for the observer looks at these entities as it were from the outside. On the other hand, basic statements are theory-laden or theory-dependent in a sense to be explained below. For the time being, suffice it to say that objective statements involve universal and hence highly conjectural hypotheses; so their truth-value cannot be determined with any degree of certainty.

[E] Accepting a basic statement thus constitutes a decision, and a conventional one at that. But the convention is not arbitrary; it rests on a consensual decision where regard is paid to the feelings of various experimenters. According to Popper,[10] a consensus concerning the acceptability of a basic statement b should be regarded as attained if there are no divergences between the feelings of certainty experienced by various observers. Popper characterizes b as being both verifiable and falsifiable, i.e. as being fully empirically decidable.[11] But it is clear from the context that these terms have little to do with the truth-value of b and could just as well be respectively replaced by 'acceptable' and

---

[10] See Popper, *BG*, p. 130.   [11] See Popper, *BG*, p. 125.

'rejectable'.[12] Thus the non-arbitrariness of the conventions adopted vis-à-vis basic statements has to do, not with their presumed truth-value, but with the uniformity of certain procedures. It is also to be noted that these procedures are partly psychologistic: the feelings of conviction of various experimenters play a crucial role in reaching a consensus as to the acceptability of a basic statement b. This is why such a consensus founds the objective, i.e. the intersubjective character of b, as distinct from its truth-value.

4.   The Structure of Popper's Argument

In what follows, clause [A] will often be referred to as *the psychologistic view of level-0 statements or, more briefly, as the psychologist thesis*. This thesis is the linchpin of Popper's argument in support of his conventionalist conception of the empirical basis. It will be shown that clauses [B]–[E] are all consequences of the psychologist thesis taken together with some trivially true principles. I shall postpone my criticism of [A] until the next section and consider now in what way the other clauses follow from [A].

Let us first remind ourselves that, according to [A], autopsychological sentences are supported exclusively by *subjective feelings* of conviction. As for [B], it is a trivially true proposition: for by definition, an autopsychological assertion q is private to one speaker and only at the instant at which he utters 'q'. As soon as an observer relies on his

[12] See Popper, *BG*, p. 127.

memory in order to describe his past experiences, his statements become fallible; for he is now dealing with events out of the reach of his present consciousness. I.e. he is now referring to strictly transcendent states-of-affairs. If we moreover accept the psychologist thesis and hence the view that level-o reports are fallible because based *only* on feelings of certainty, then it becomes preferable to look for other kinds of 'basic' statements; that is, if we are to be in a position to test our hypotheses severely.

Clause [C] also follows from [A]. Psychology does investigate feelings of conviction and will clearly face a problem of vicious circularity, should it decide to found *its* methodology on basic statements underwritten exclusively by such feelings.

Finally, [D] is also a consequence, albeit an indirect one, of the psychologist thesis. For a statement is either immanent or transcendent; where, to repeat, 'transcendent' means: making at least one reference to some entity external to our mental operations. Should immanent propositions prove unsuitable for the role of basic statements, we would thus have to resort to sentences some of whose referents are mind-independent, i.e. to transcendent propositions. We now have to turn to the problems posed by the alleged theory-ladenness of such propositions.

5.   In what sense or senses are observation statements theory-dependent?

Duhem was among the first epistomologists to underline the theory-dependence of all scientific propositions, which he sharply distinguished from singular commonsense statements.

> An experiment in physics is the precise observation of phenomena accompanied by an *interpretation* of these phenomena; this interpretation substitutes for the concrete data really gathered by observation abstract and symbolic representations which correspond to them by virtue of the theories admitted by the observer. . .
>
> It is not correct to say that the words 'the current is on' are simply a conventional manner of expressing the fact that the magnetized little bar of the galvanometer has deviated. . . This group of words does not seem to express therefore in a technical and conventional language a certain concrete fact; as a symbolic formula, it has no meaning for one who is ignorant of physical theories; but for one who knows these theories, it can be translated into concrete facts in an infinity of different ways, *because all these disparate facts admit the same theoretical interpretation. . .*
>
> Between the phenomena really observed in the course of an experiment and the result formulated by the physicist there is interpolated a very complex intellectual elaboration which substitutes for the recital of concrete facts an abstract and symbolic judgement.[13]

This long passage was quoted for two reasons. First, it provides a synopsis of Duhem's position on the status of factual scientific propositions. It secondly hints at two different, not to say at two disparate notions of theory-ladenness;

[13] Duhem, *The Aim and Structure of Physical Theory*, Chapter 4, Section 1.

the one is objective-logical, the other psychological. But more of this later. Let us now note that, according to Duhem, particular commonsense claims like 'There is a white horse in the street' possess truth-values which can be infallibly determined. Such statements do not however belong to science proper. As for the descriptions of genuinely scientific facts, they are precise, symbolic and theory-laden. At times, Duhem seems on the point of denying any determinate truth-value to such propositions; at others, to hold that their truth-value cannot be effectively determined. Be it as it may, an increase in scientific precision is paid for by some loss of certainty; and there exists, between science and commonsense, a hiatus which contemporary French thinkers call 'coupure épistémologique'.

Popper broadly shared Duhem's view of the empirical basis; but he rightly underlined the continuity between science and commonsense ('Science is commonsense writ large'), hence the fallibility of both types of knowledge. 'There exists a white horse in the street' and even 'I see a white horse in the street' can both be false; for I may be hallucinating. Even if my perception were 'veridical', the concept of horseness would still be theory-dependent: it rests on hypothetical assumptions, no matter how minimal, about the existence of animal species.

Let us return to the concept of theory-ladenness, which has recently been overworked and is therefore in need of clarification. As indicated in the above quotation, theory-ladenness is sometimes equated with the notion that the meanings of observational terms are theory-dependent. This thesis seems to me totally unconvincing. If a scientific system is to entail testable consequences, it must contain at least some observational terms, whether these be primitive or defined. According to usual Tarskian semantics, the truth-value of the system is founded on the meanings or rather on the referents of its primitive concepts; *not* the other way round. It is thus difficult to see how the meaning of an observational term can presuppose the truth of a theoretical premise or, for that matter, of any premise whatever. Saying that the meanings of terms are theory-dependent sounds like putting the cart before the horse; unless by meaning one meant something like a verbal or dictionary definition, whereby a single word is treated as shorthand for a much longer description formulated in terms of an underlying primitive vocabulary.

This brings me to the central point of this section; the point namely that in the thesis of theory-ladenness two distinct propositions are often conflated, namely: (a) a logical principle according to which the so-called observational terms are definite descriptions containing occurrences of universal theoretical sentences, and (b) the cognitive-psychological thesis that what we observe depends, not directly on any theory, but indirectly on our believing or even on our entertaining or understanding certain hypotheses.

Let us start with the logical point. The sentence 'The current is on' looks like a singular proposition because 'the current' is instinctively treated as a proper name; but it actually represents a definite description which, when properly unpacked, is found to involve a sophisticated electrical theory. For example, 'the current' could stand for: the

movement of certain electrons in a well-defined spatio-
temporal region, where the electrons are subject to electro-
dynamic forces. Similarly, in Popper's favourite example
'Here is a glass of water', 'water' stands for something like:
a substance resulting from the combination of one oxygen
and two hydrogen atoms, where the combination is effected
by certain cohesive forces. A weak parallel can be drawn
between these basic scientific statements and 'The present
king of France is bald', which stands for: there exists exactly
one present king of France and, for all x, if x is any present
king of France, then x is bald. The longer sentence contains
quantification over individual variables and is therefore
non-singular (non-atomic). Needless to say, the universal
assumptions involved in genuinely scientific factual propositions are more complex than those in 'The present king of
France is bald'. Thus, basic statements possess only a *superficially singular syntax* but a *deep universal-theoretic* structure. In this sense, they are theory-dependent and hence
fallible. Popper himself admits that his basic statements
constitute hypotheses, albeit of the low-level sort. Their
theoretical structure enables them to be further tested: they
can be adjoined to other hypotheses and thus give rise to
new predictions, over whose truth-value the scientific community may find it easier to reach agreement. Even the
adjective 'low-level' seems to have little cash-value: in the
examples above, the assumptions involved in basic statements are among the most fundamental laws we possess.
This is why I have to differ from John Watkins's view that
hypotheses can be tested, hence falsified, by level-1 propositions. The basic statements mentioned above seem, at least

intuitively, to belong to a level far higher than the first one. So all we have before us is a logical conflict between two theories whose levels are on a par. And only our attitude decides which hypothesis is under test and which does the testing. In epistemology, the only legitimate divide appears to be the one separating level-o reports from the rest, i.e. from all transcendent propositions.

Let us now turn to the psychological thesis (b). It says that observation largely depends on the theories held to be true or even merely entertained by the observer; the latter for example jumps from some visual experience, of whose intricate details he often remains unaware, to a description of the assumed cause of his perceptions. Such a cause is inferred against the background of theoretical assumptions instinctively made by the observer. An experimenter will thus observe different things depending on his theoretical prejudices. Schopenhauer rightly maintained that even animals make instinctive use of an inborn principle of causality in order to adjust to their surroundings. He should have added that, in these cases, the inferred causes are always the same under the same circumstances; humans seem to be more flexible, hence more adaptable than other animals. Note further that even if one insisted on speaking of sense-data as the proper objects of perception, such private entities are clearly far from being passively given: each possesses a focus towards which our attention is directed by our pre-conceived ideas. Thus both focus and attention depend on our beliefs and, more generally, on the overall state of our knowledge. In other words, a sense-datum is structured largely by the mental acts which intend it.

In the above quotation, Duhem speaks of a scientist *interpreting* his experiences in the light of some theory. 'Interpretation' seems to me to be a misnomer; we do not *interpret* our perceptions but immediately *infer their presumed causes*, where the latter depend to a large extent on our background knowledge. Let us give a few examples. Michelson claimed to have ascertained ether drag. He took it for granted that light was a wave phenomenon presupposing a carrier called 'ether'. This medium was either dragged, i.e. pulled along by the motion of the earth, or there was slippage between earth and ether. This is why Michelson could not help 'interpreting' his null result as directly revealing ether drag. Nowadays, we have good reasons for supposing that no ether exists; so Michelson could not have observed any medium being dragged by the earth's motion. His experimental claim is false, thus establishing Duhem's and Popper's theses about the fallibility of basic statements.

Galileo's discovery of Jupiter's moons provides another illustration of the theory-dependence of experimental results. In his famous play, Bertolt Brecht claimed that when asked to look through the telescope and thus observe these moons, Galileo's opponents refused point blank to carry out this simple experiment; they had allegedly decided a priori that such objects as the Jupiter moons could not exist. This story may well be apocryphal; but even if they had looked through the telescope, Galileo's critics might not have been convinced; they would probably have either failed to focus on the four luminous spots or else 'interpreted' them in a way very different from Galileo's. I.e. they would have told a different causal story; given their Aristotelian background, they might have claimed that the spots emanated from the telescope itself. But Galileo himself had prejudices, albeit Copernican ones: because he believed the world to be polycentric, he jumped to the conclusion that he was observing satellites of Jupiter.

Thus beliefs in different theoretical systems give rise to incompatible *transcendent* basic statements. Could we not however maintain that in all these cases, though there are major differences in the 'interpretations' of the facts, i.e. in the way some causal account is given, there exists a fixed perceptual core which does not depend on the experimenter's theoretical prejudices? After all, Michelson must have seen the *same* interference fringes which a modern relativist would observe, were he to repeat Michelson's experiment. And would not Galileo's enemies have seen the same four spots which he claimed to have observed?

In answer to these well-known objections, let us first note that we have already shifted from speaking about physical entities to describing certain experiences, that is, we are moving towards a more phenomenological level. Even then, sameness of perceptual content cannot always be maintained. We have already remarked that focus and attention are largely determined by preconceived ideas. On a chest X-ray a trained doctor will notice details which escape the attention of the layman. The latter will—at best—recognize the normal contours of the patient's ribs amid areas of varying shades of grey; whereas the physician's attention will be drawn to a dubious patch indicating the presence of a tumour or of some infectious disease. Even if we restrict ourselves to a phenomenological language, we still have to say that the doctor and the layman see *different phenomenal objects*: the doctor, but not the layman, homes in on the 'diseased' patch which forms a central point of his sense-datum. In as far as focus and attention form essential parts of every intentional entity, the two sense-data, the layman's and the physician's, are different. As for the real objects which are said either to be observed or to be immediately inferred—namely the diseased as opposed to the allegedly normal lung—these are clearly different in the two cases.

To sum up: in all the cases cited above, we mistakenly refer to the *theory*-dependence of observation reports; whereas we ought more strictly to speak of the dependence of experimental results on the experimenter's *beliefs* and, more generally, on his *state of mind*.

This point seems pedantic and somewhat trivial; but it underlines the crucial role played by psychological assumptions in the derivation of basic statements (or more exactly: of their negations). This brings physics into line with the social sciences, where it was long realized that the prediction of some phenomena may depend on the subject of an experiment being aware of certain theories, more particularly of the hypothesis under test. Important methodological consequences flow from this 'theory-dependence' of observational results; consequences which are grist to my phenomenological mill. We have seen that Michelson's claim to have observed ether drag is false. Yet the basic statement 'There exists ether drag' (at some point of the earth's surface), was taken to have refuted classical electrodynamics, which is now also thought to be false. We have here a clash between two propositions which are not only fallible, but both of which are actually false. So there was no reason for us to conclude that classical electrodynamics had been refuted. I.e. we were merely lucky in rejecting a false hypothesis but had no really good reason for doing so. Our decision becomes much more rational if we switch attention to what ought to have been Michelson's more guarded, *autopsychological* claim: 'I seem to have detected ether drag' which we still take to be true—of Michelson's state of mind. But in order to deduce such statements or their negations from classical electromagnetism, we have to conjoin to the latter some auxiliary hypotheses including assumptions about the experimenter's beliefs. In this way we obtain, as premises of our derivation of phenomenological basic statements, a greatly expanded scientific system consisting not only of some *core physical theory*, but also of psychological and of psycho-physical hypotheses. We must furthermore resort to descriptions of boundary conditions and of the experimenter's mental state. Needless to say, this increases the magnitude of the Duhem–Quine problem; but, in return, we can now at least know or rather Michelson would have known, that his premises, taken as a whole, had been falsified by the ether *seeming* to him to have been dragged.

6. Consequences of [B]–[E] for Popper's position

In the last few paragraphs I seem to have jumped the gun. Going back to Popper, we realize that his position becomes perfectly coherent once we accept clause [A], i.e. his psychologistic conception of level-o reports. For we have seen that [A] entails [B], [C] and [D]. As for [E], it merely expresses the *stipulation* that basic statements should be accepted on the basis of a consensus. Hence [E] does not have to follow from *any other proposition*. The other clauses nonetheless provide a rationale for adopting [E].

We shall now show that [B]–[D] have serious consequences for Popper's overall position which, contrary to his own intentions, is thereby turned into out-and-out scepticism. Throughout this section and barring indications to the contrary, the psychologist thesis will be presupposed. Basic statements—also called potential falsifiers—are taken to be objective *theory-laden* propositions; theory-laden both in the logical and in the psychological senses described above. Potential falsifiers are therefore doubly fallible and can thus be accepted only through a consensus among experimenters. In Popper's own words, basic statements constitute dogmas, albeit temporary ones. Such dogmas are said to be harmless because they can always be revized; being in the nature of low-level hypotheses, they can be added to other assumptions and hence further tested. But by what? By further potential falsifiers which, in their turn, are to be accepted through the fiat of some consensus. At no point does this potentially infinite process involve genuinely epistemological considerations; i.e. *considerations which might link the actual or presumed truth-value of a proposition to an individual act of observation or to the procedure of a consensus.*

According to Popper, traditional conventionalism and his own brand of empiricism differ mainly over the type of statement which both conventionally accept.[14] Conventionalism holds on to theories while Popper gives methodological priority to the potential falsifiers of such theories. This provides an illuminating criterion for demarcating between conventionalism and Popperian empiricism, though hardly any reason for preferring the latter to the former. In fact, Popper's view of the empirical basis threatens to destroy the presumed asymmetry between verification and falsification. For we now face a situation in which one hypothesis which we label 'theory H' confronts another hypothesis, which we label 'basic statement B'. But all we are entitled to assert is that H proves logically incompatible with B; so the relationship between H and B is perfectly symmetrical. As already mentioned, Popper describes B as a *low-level* hypothesis, but this appellation is of no great help; for since neither of the two propositions, H and B, logically implies the other, the levels of H and B are strictly non-comparable. More seriously: as explained above, nothing tells us that the theories impregnating B are less risky than H; yet we arbitrarily regard H rather than B as being under test. All we can say is that over the last 400 years or so, science has pursued a largely empiricist policy; i.e. physicists have reached agreement over statements which have *superficially* the same form as B rather than over statements *resembling* H. But as Popper himself admitted in 'Die beiden Grundprobleme', we can offer no scientific explanation for the success of this empiricist strategy.[15]

---

[14] See Popper, *BG*, pp. 129–130.   [15] See Popper; *BG*, p. 132.

Worse still: we cannot explain why the pursuit of such a policy since the beginning of the scientific revolution issued in undeniable *technological breakthroughs*. Such breakthroughs are taken to depend at least on the approximate truth of some consequences of our theories. Yet truth considerations have so far played no role in Popper's methodology: in rejecting H in favour of B, we have no good reason for supposing that we have thereby eliminated error, let alone that we have approximated truth. Sustained technological progress thus becomes a perpetual miracle.

As so far depicted, Popper's position clearly leads to scepticism about every item of knowledge save logic and mathematics. And the only criticism we can level at a universal theory H is of the transcendent type, which Popper had firmly rejected in the 'Die beiden Grundprobleme':[16] we simply pit against H another assumption B which is logically incompatible with H.

As is well known, Popper's fallibilist attitude towards the empirical basis provided the late Paul Feyerabend with his major argument in favour of epistemological anarchism. As set out in 'Against Method': if Galilean theory was supported only by *Galilean experimental results* while Aristotelianism rested on *Aristotelian observation*, then Galileo could and did win only a propaganda war against the Aristotelians. The observation reports of neither school could be accepted by the other. Hence *anything goes*,

---

[16] See Popper, *BG*, p. 53. Unfortunately, Popper changed his mind—but only partially—as to the inadmissibility of transcendent criticism. Still, he regarded the latter as insufficient to refute the criticized position.

provided it be defended by means of a powerful and persuasive rhetoric.

It has been shown that, according to Popper, a level-o approach to the empirical basis would lead us straight into psychologism. But Popper also recognized the clearly socio-logical character of his own conventionalist view. Let us recall that his concept of objectivity is founded on the twin conditions of repeatability and consensus; this means that science is entitled to ignore observation reports which cannot be intersubjectively ascertained.[17] But would it not then follow that descriptions of dreams cannot provide tests for psychological hypotheses? A dream is an intrinsically private phenomenon whose repetition, even in the case of one and the same person, seems to pose insuperable prob-lems. Admittedly, certain aspects of a dream might recur, but one can hardly speak of repetitions in the sense in which a physical experiment can be rerun in a controlled way. As for the question of intersubjectivity, it is difficult to see how it could even arise in the case of purely subjective phenom-ena. For example, in a laboratory, any number of technicians can look through the same microscope at the same blood sample and arrive at the same conclusions; but it would clearly be absurd to ask these technicians to 'observe' or experience the same dream. Yet Popper would surely not want a priori to deny scientific status to psychology.

There is another objection to the conventionalist thesis which is also mentioned by Popper in 'Die beiden

---

[17] See Popper, *BG*, p. 122 (my translation).

Grundprobleme', namely the 'Robinson Crusoe Objection':
it follows from Popper's demarcation criterion that no
person on his own, e.g. if stranded on a desert island, can
do any science.

> One can imagine a person, a Robinson Crusoe, who
> though completely isolated, would master a language and
> develop some physical theory (possibly in order to gain
> better mastery over nature). One can imagine—though
> this is far from being plausible from the viewpoint of the
> psychology of knowledge—that his physics coincides, so
> to speak literally, with our modern physics; furthermore
> that such a physics was experimentally tested by Crusoe
> who has built himself a laboratory. Such a process, no
> matter how improbable, is at any rate thinkable.
> Consequently, so concludes the Crusoe objection, the
> sociological aspect is of no fundamental significance for
> science...
>
> With respect to this argument, one must concede
> that continued testing by one individual is already
> similar to intersubjective testing (the sociological aspect
> is thus, at any rate in many cases, of no decisive
> significance for science). Further, the concept of
> intersubjectivity, of the multitudes of subjects, is in
> certain respects not sharply defined. Yet the Crusoe
> objection does not hold water: that which Crusoe
> constructs under the name of physics is not a science;
> this is so not because we arbitrarily define science in
> such a way that only intersubjectively testable theories
> can be termed scientific, but because the Crusoe
> objection starts from the false premise that science is
> characterized by its results rather than by its
> methods... (BG, p. 142, my translation)

To my mind, Popper's argumentation betrays a certain embarrassment. Let us first remark that Robinson Crusoe need not invent sophisticated hypotheses comparable to those of modern physics; he may construct simple laws about regularities in his physical environment, then use these laws in order to survive. He would thereby be implicitly testing his theory and hopefully surviving the successive tests. Thus, Crusoe's discoveries and success need not be miraculous. Should he regrettably succumb as a result of his first test, then his hypothesis would certainly qualify as scientific; for it would have been refuted, thus sadly preventing him from repeating the experiment. This conclusion holds good even in cases where the test can *in principle* be performed only once. Hence, even the possibility of having an experiment *repeated* by the same person constitutes no *essential* requirement. Notice secondly that, in the passage above, Popper makes the concession of sometimes equating intersubjectivity with the repeatability of the same test by the same person. He then annuls his concession by claiming that the Crusoe objection is anyway invalid because it mistakenly takes science to be defined by its results and not by its methods: according to Popper, Crusoe does not practise science because he does not apply any genuinely scientific method. But this is precisely where the problem lies: if we build repeatability and intersubjectivity into the very definition of basic statements, then Robinson Crusoe can, by definition, no longer pursue any science.

Let us again recall that, should we accept the psychologist thesis, then we should also have to accept clauses [B]–[E], hence all the counterintuitive consequences just mentioned. But these consequences *are* intuitively very hard to take on board. After all, as its name indicates, physical science is founded on man's relation to physical nature which acts on man through his senses and his brain, thus finally reaching his consciousness. Only at this level does man's contact with nature occur unmediated, which is why most scientists subscribe to the phenomenological and not to the consensual view of observation. To the extent that Crusoe tries out his hypotheses, then corrects and uses them in mastering his physical environment, he ought to be practising empirical science. He does not of course possess as efficient a method for eradicating error as we do. If he is colour-blind, he may well perish in circumstances under which we should have a greater chance of surviving; but he could conceivably develop a theory about frequencies enabling him indirectly to distinguish between all 'colours'. He is in greater need of luck than we are, but we collectively need a lot of luck too. Thus social conditions may provide additional means of detecting error, i.e. of unravelling the Duhem–Quine problem; but they are not of the essence. Methodologies should anyway remain neutral vis-à-vis such contingencies.

Now consider one last difficulty which flows from the psychologist thesis i.e. from clause [A] above. [C] tells us that psychology would face a problem of vicious circularity if it decided to base its methodology on the observers' feelings of conviction with respect to basic statements. This argument is undoubtedly valid, but we can see straight away that sociology fares just as badly under Popper's conventionalist thesis. Sociology investigates how certain consensuses come about and whether they are due to purely ideological or social factors; or whether they are at least partially determined by objective considerations; e.g. by theorizing about the physical world in cases where the consensus is about some alleged physical fact like the earth's immobility. Sociology may moreover examine whether and under what conditions different consensual agreements are mutually consistent. According to Popper's conventionalist view of basic statements, a sociological hypothesis ought to be tested by reports underwritten by a consensus. Of course, a sociological theory might tell us that the consensus is objectively well-founded; but this theory may well be the one undergoing the test, so that its status is highly uncertain. Or else it might claim that only superstitious beliefs led to the consensus in question. In neither case can we make use of the allegedly factual reports in order either to refute or to confirm the hypothesis under test. Popper was aware of this paradox when he wrote:

> Knowledge is thus possible because there are
> 'unproblematic' basic statements (an analogue to the
> intuitively certain propositions about sense-data), i.e.
> basic statements which need not be tested any further
> and which should not be questioned after intersubjective
> agreement has been reached. That such propositions can
> exist, that we have had luck with such decisions and with
> such experimental propositions, that we are not thereby
> led into contradictions is to be noted as a fundamental
> methodological fact; a fact of which we can naturally
> never know that it will obtain at all times and in all cases.
> (Why there are such propositions, why objections are not
> raised against every decision or why the decisions do not

lead to contradictions, this *question*, like all questions
about the grounds of the possibility of knowledge is
scientifically impermissible and leads to metaphysics, to
metaphysical realism and not to a realism of method.)
(*BG*, p. 123, my translation)

Despite the coherence of Popper's position, I find
his conclusions puzzling. There ought to be no good reason
for excluding a priori the possibility for *science* of explaining
both the occurrence and the consistency of intersubjective
agreements in terms, say, of the observers' shared biology
and social environment. Anyway, should we succeed in
refuting the psychologist thesis, then the phenomenological
view will—in principle—give us a free hand in explaining
the occurrence of consensual agreements and of feelings of
certainty with regard to experimental propositions.

That Popper later recognized the problematic character of his conventionalism with respect to the empirical
basis is shown by his reply to Ayer's criticism. According to
Ayer, subjective experience provides not only a motive or a
cause, but also a reason for accepting objective experimental
results:

> None of this prevents it from being true that my having
> this 'observational experience' supplies me not only with
> a motive but also with a ground for accepting the
> interpretation which I put on it...
>
> It is to be noted that in arguing that basic statements
> can find a justification in experience, I have not found it
> necessary to cast them in the form of statements which
> refer exclusively to present sense-data. This is not
> because I have any objection to such statements in

principle; I am not even convinced that it is wrong to
treat them as incorrigible, in the sense that they do not
leave room for anything other than merely verbal
mistakes. It is rather that I do not think that they are
strictly needed for the role which has here been assigned
to them. ('RC', p. 689)

My main objection to Ayer's argument is that it tries to
circumvent all recourse to level-o sentences by claiming
that experience directly confirms objective basic statements. But a proposition is justified either by the existence
of the state-of-affairs it denotes or through the medium of
other propositions. Popperian basic statements are about
public events, hence cannot not have private experiential
referents. The latter correspond to level-o reports, i.e. to
Ayer's sense-datum statements. We must therefore take
Ayer to be effectively saying that perceptions verify sense-
datum propositions, which in turn justify some 'interpretations' of our experience, i.e. some Popperian basic statements. By 'justification' Ayer clearly means something like
inductive support, which falls short of full verification.
Given his global anti-inductivist stance, Popper could
consistently have rejected Ayer's thesis. But it also seems
clear that, short of admitting to full-blown scepticism,
Popper ought to have accepted sense-datum statements
as explananda within a hypothetico-deductive scheme.
That is I think that Popper should have looked upon
sense-datum reports as basic statements or, more precisely, as negations of basic statements to be deduced from
an expanded scientific system. Instead of making this
move which would have been consistent both with his

falliblism and with his anti-inductivism, Popper chose to cave in to Ayer's argument.

> Nevertheless, most organisms act upon interpretations of the information which they receive from their environment; and the fact that they survive for some considerable time shows that this apparatus usually works well. But it is far from perfect...
>
> ...Our experiences are not only motives for accepting or rejecting an observational statement, but they may even be described as inconclusive reasons. They are reasons because of the generally reliable character of our observations; they are inconclusive because our fallibility. ('RC', pp. 1112–1114)

Popper's agreement with Ayer that observation supplies us with (inconclusive) reasons for accepting basic statements thus rests on the following argument. External states-of-affairs causally give rise to some of our experiences; these in turn motivate us into believing certain basic statements; the latter are not however about the experiences themselves, but about their external causes. So far, this chain is a purely causal one, where we directly respond to the *external* world because of our need quickly to adjust to it in order to survive. At this point, a Darwinian twist to the argument takes place: in many if not in most cases, natural selection sees to it that our basic statements correspond to, or at least approximately model their transcendent causes, which thus become their referents; otherwise, we should not have survived long enough to start the present philosophical discussion. Like Ayer, Popper obviates all recourse to level-0 reports, but at very high cost to his own anti-inductivist position. This is not to say that Popper's claims are false, only that they presuppose Darwinism; they are circular and thus constitute no *argument* for the reliability of basic statements. If we happen to think that Darwinism is largely correct, this is only because its broad aspects have been strongly corroborated. The observation reports supporting Darwinian theory should not however be predicated on Darwinism itself. In 'Die beiden Grundprobleme' Popper realized the necessity of avoiding vicious circularity; he agreed with Leonard Nelson that a *theory* of knowledge, conceived as a system of true-or-false statements, is impossible; for it leads either to circularity or to an infinite regress.[18] This is precisely why Popper intended his methodology to consist of a set of proposals rather than propositions. For no *transcendent* statement can be appealed to prior to fixing some methodology, since any argument in support of such a statement involves theories about the external world whose confirmation presupposes a methodology.

Both Ayer's and Popper's theses are vitiated by their refusal to take account of the special status of autopsychological reports. These are based not on any *science* in the ordinary sense, but on phenomenological analysis. And phenomenology, whose founders intended its principles to yield those of logic and of mathematics, may well prove to be the last, unavoidable and irreducible bastion of the *synthetic a priori*.

[18] See Popper, *BG*, p. 110.

7. Refutation of the Psychologist Thesis

It has to be admitted that Popper had sound historical reasons for holding a psychologistic view of autopsychological reports. This last phrase anyway has a pleonastic ring to it. As explained in the chapter of 'Die beiden Grundprobleme' which is devoted to Kant and Fries, Fries realized that, on pain of falling into vicious circularity or into some infinite regress, we have to give up the idea of justifying *all* synthetic propositions. In order to avoid dogmatism, Fries accepted experiential statements as something which, as a matter of brute psychology, we cannot help believing in. Thus Fries openly subscribed to the psychologist thesis.

Franz Brentano, a compatriot and near-contemporary of Popper, tried to account for the certainty of level-o reports in non-psychological terms. He thereby developed an 'Evidenzlehre',[19] i.e. a theory of self-evidence. Brentano distinguished between the blind compulsion which misleads us, even in cases of hallucination, into believing in the existence both of allegedly external objects and of their secondary qualities; and genuine self-evidence which is the objective hallmark of indubitable propositions, e.g. of the logical and mathematical principles, of the axioms of probability and of some autopsychological statements (which he called statements of secondary or inner perception). The latter are however known to be contingent; so they cannot be established without appealing to some synthetic principle. Once

[19] F. Brentano, *Wahrheit und Evidenz*, Section 4 (F. Meiner, 1930).

again, the twin problems of infinite regress and of vicious circularity rear their ugly heads. Brentano envisaged two solutions to these problems. The first consisted in giving a mere laundry list of propositions which were, by fiat, declared to be self-evident.[20] This is tantamount to postulating what is needed and, as well-known, represents the advantages of theft over honest toil. According to the second 'solution', self-evidence can be infallibly recognized through direct insight.[21] Although self-evidence was regarded as an objective attribute of certain propositions, one had in the last analysis to rely on a psychologistic criterion in order to identify it. Popper could therefore be excused for taking such 'Evidenzlehren' to be thinly disguised forms of psychologism. However, I think that Brentano's thesis can be rescued by methods somewhat similar to those used by one of his followers, Anton Marty.

As already mentioned, in trying to found the truth of level-0 propositions, one has to stop somewhere; and this 'somewhere' must be synthetic, hence questionable. So why not stop, as Brentano did, at the level-0 reports themselves? One could simply dub such reports self-evident. But where one breaks off the analysis is not a matter of indifference. In the present case we may be able to determine, for the certainty of autopsychological statements, a source which links up with other components of Popper's philosophical system. Although some unquestioned principle must in the end be invoked, this same principle may be needed in other

[20] Ibid. Section 4.   [21] Ibid. Einleitung.

areas within a global position, which thereby becomes more unified.

In what follows, I propose to show that the certainty of level-0 sentences rests on an intuitive version of the Correspondence Theory of Truth; and it is significant that Brentano resorted to his 'Evidenzlehre' precisely at the time when he expressed doubts about the Correspondence Theory. Remembering that Truth-as-Correspondence became a central theme in Popper's later philosophy, my move will demonstrate that the phenomenological view of the empirical basis forms a natural complement to Popper's methodology.

Putting it briefly, I shall show that level-0 reports can be recognized as true not by being derived from self-evident propositions, but through our having a direct access both to their referents and to the meanings of all concepts occurring in them. Only in such epistemologically privileged situations can we ascertain whether a truth-correspondence obtains between a sentence and the state-of-affairs it is supposed to signify. Now, we have to analyse what kind of referents autopsychological propositions can be said to possess.

Ayer wrote:

> The next step, continuing with our example, is to convert the sentence 'it now seems to me that I see a cigarette case' into 'I am now seeing a seeming-cigarette-case'. And this seeming-cigarette-case, which lives only in my present experience, is an example of a sense-datum ...
>
> ...the expression must be used in such a way that the experience of the physical object which appears to be

referred to remains an open question: there is no
implication either that it does exist or that it does not...

What appears most dubious of all is the final step by
which we are to pass from 'it seems to me that I perceive
x' to 'I perceive a seeming-x', with the implication that
there is a seeming-x which I perceive...

... For to talk of someone's sensing a sense-datum is
intended to be another way of saying that he is sensitively
affected; the manner in which he is affected reappears as
a property of the sense-datum ...[22]

In these passages Ayer mentions, as if in passing, one of the
central problems with which Brentano struggled throughout
his life. Brentano underlined the intentional structure of
consciousness which, in his view, is always a consciousness
of something. Every mental act intends, i.e. it is directed
towards, some object or other. Such objects need not exist
independently of the mental operations which generate
them; they were consequently called 'intentional' or 'immanent'. This does not mean that no transcendent entity exists
over and above the intentional one; only that the latter is
always present, whether or not the former is. Given the
phenomenological continuity between hallucination and
veridical perception, all perceptual objects must belong to
the same category and are therefore non-physical. In view of
the above quotation, Brentano's intentional objects of perception can be identified with Ayer's sense-data. But can
immanent entities be said to exist in any legitimate sense?

[22] A. J. Ayer, *The Problem of Knowledge*, pp. 106, 111, 115, 118 (Penguin,
1956).

(Note that the answer must be in the affirmative if 'I see a seeming-x' is to count as a true level-o report.)

Initially, Brentano ascribed to intentional entities a half-hearted kind of existence which he misleadingly called 'inexistence in consciousness'. Sense-data were not supposed to exist the ordinary realist sense but to remain private to one mind; still, they were *things* which could act as bearers of such phenomenal properties as colour, smell and shape. They were interposed between the mental acts on the one hand, and the external world on the other; they could in principle reflect the properties of transcendent objects, but they existed only 'in the mind'. Should this thesis prove tenable, then we could identify level-o statements with propositions of the form 'I perceive a seeming-x', where x is a definite description. The referent of 'seeming-x' will be some sense-datum. Since the latter is an immanent entity, we can, if this entity exists, directly apprehend it and check whether it has the phenomenal properties expressed by x. Level-o statements would thus have ascertainable truth-values.

With time, Brentano grew dissatisfied with his shadowy intentional realm; and this for a variety of reasons. To begin with, he became more of an empiricist, hence doubted the existence of such ethereal entities as sense-data. Later, he also felt that the postulation of objects of thought would inevitably engender logical paradoxes. He was afraid of being driven into accepting the existence of logically impossible objects like round squares. In this, he was undoubtedly mistaken: there is of course nothing to stop us conjoining squareness to roundness, thus obtaining the empty class; but we need neither perceive nor even imagine round squares.

Hence the principle of non-contradiction is not threatened by the introduction of objects of thought.

All the same, Brentano seems to have had an unerring instinct regarding the logical inadmissibility of intentional entities; for these are both richer and less determinate than ordinary objects. They are richer in that they present themselves theory-laden; thus, the same physical object can give rise to infinitely many different sense-data: what was said above concerning the theory-ladenness of Popperian basic statements can be transferred to all descriptions of (primary) intentional entities. We have already noted that sense-data have a focus which is fixed by the observer's attention; so that the same chest X-ray can be integrated in infinitely many ways. Two-dimensional objects, e.g. a painting or an image on the television screen, are automatically observed as possessing depth. Only through an effort of reflection can I bring myself to 'see' them as two-dimensional pictures. The photo of an object taken from an unusual angle will, on the contrary, appear as a two-dimensional arrangement of colours and shapes; the object may then be recognized and the new sense-datum will not only supplant but also suppress the old one. In the case of a dot on a radar screen, both the expert and the layman will be aware of a two-dimensional picture, to which only the expert will instinctively add some information about the distance of a flying object and only through an act of will can he suppress this theoretical admixture and 'retrieve' a purely visual image. Since every transcendent object may give rise to any number of sense-data, admitting the existence of the latter could lead to ontological hyperinflation. But there are worse dangers still; for intentional entities are less determinate than their physical counterparts: they possess only those properties which are explicitly projected on to them by our mental acts. A sense-datum is after all a phenomenal entity, i.e. some *object-as-perceived*. A speckled hen is apprehended as having a finite number of speckles but neither as having 100 speckles nor as not having 100 speckles. Should we transform this innocuous sentence into a statement about sense-data, then we would clearly violate the law of the excluded middle: the phenomenal hen will have a finite number of speckles but it is not the case that this number will be either equal to or different from 100. In this respect, the hen is on a par with Hamlet who cannot be said to be 1.75 metres tall or not 1.75 metres tall. Hamlet was simply not intended to have any specifiable height. That is, an intentional entity will lack every property which is not explicitly attributed to it by some act of consciousness: Husserl expressed this principle by asserting the existence of a strict correlation between the noesis on the one hand, and the noema together with its intentional object on the other. I take this to mean that all talk of intentional entities, and more particularly of sense-data, is shorthand for some description of the underlying mental operations.

One can of course repair this situation by decreeing that even sense-data obey the law of the excluded middle; but they will then escape our control and cease to be immanent entities. For we may be unable to count the number of speckles on the 'phenomenal hen'; and any mechanical procedure for counting them will alter the corresponding sense-datum. Finally, should the hen have been hallucinated, then to assume that it would still possess a well-defined number of speckles is nothing short of fantastic. Hence, for any intentional object y and any property P which y might possess: P(y) holds iff y is intended, by the corresponding noesis, to have P.

Brentano was thus right to look for an alternative solution, which he put forward at the turn of the century and which he held on to until his death in 1916. Before discussing Brentano's thesis in any detail, let us examine a concrete example. Consider the proposition 'I see a red house' and assume that there is no physical house out there at all; that is suppose that I am hallucinating. If 'red house' is given its ordinary realist meaning, then my proposition will clearly be false. It could conceivably be turned into a true sentence if 'red house' were to denote a sense-datum. But as already explained, we have good reasons for denying the existence of all intentional entities; so the truth-value of my proposition becomes problematic. Fortunately, the sentence 'I seem to be perceiving a red house' has a truth-value which is independent of the existence or non-existence of the referent of 'red house'. And this is all we need in order to secure a phenomenological basis for the sciences.

According to Brentano's definitive solution, all forms of consciousness are described by statements of the form 'I seem to be perceiving a red house'. That is, consciousness always possesses two indissolubly connected components (it is '*zweistrahlig*': it consists of two rays). There is a first primary conscious act directed at a primary object, e.g. the red house. However, in no sense can a primary intentional object like the red house, *qua* sense-datum, be said to exist; there simply is no immanent red house. Only the ego with its various directional activities exists. All statements about the phenomenal red house are shorthand for descriptions of modes of the underlying conscious activity. Secondly, the primary act is always accompanied by a so-called inner or secondary consciousness, which takes the primary conscious activity for its immanent object; such an immanent entity exists in the ordinary sense and, through it, secondary consciousness obliquely intends the non-existent primary object, for example, the red house *qua* sense-datum. That is: whenever I primarily seem to be perceiving a house which is red, I am at the same time secondarily aware of so seeing it. Finally, this inner awareness is aware of itself as well as of primary consciousness. Thus according to Brentano's definitive view, every conscious act is such that:

(1) There exists a primary conscious activity directed at a nonexistent primary intentional object (which does not of course imply that no corresponding transcendent object exists).
(2) There is also a secondary conscious activity which directly targets the (existent) primary act and, through the latter, obliquely or indirectly refers to the (non-existent) primary object.
(3) Secondary consciousness is finally aware of itself. Hence, tertiary and all higher levels of consciousness collapse back onto the secondary level.

There is something wrong with Brentano's final solution, but I think that it can be altered into an acceptable thesis. Note first that level-0 reports should be put in the form 'I seem to be perceiving x' rather than 'I perceive a seeming-x'. We are thus dealing with descriptions—issued by secondary consciousness—of primary mental acts; we do not therefore have to become entangled in any discussion about the ontology of sense-data, that is, of seeming-x's. Secondly however, we find Brentano defending the view that *every* mental act must be accompanied by secondary consciousness and should therefore be self-aware. I find this thesis unconvincing. Not only does it exclude subconscious activities; it also entails that animals must either be capable of self-consciousness, or else have no consciousness at all. There is anyway direct evidence—supplied by memory— that in everyday life we are not always reflexively conscious of perceiving something; we do not always observe ourselves observing; we often merely perceive; that is, we intend exclusively *primary* objects. But it is also true that the primary acts themselves can, through reflection, be made into the existent objects of higher mental activities, for example of inner or secondary consciousness. This gives rise to the level-0 reports 'I seem to be seeing a red house' and 'I believe there is a red house in front of me', whose intentional objects, namely my seeing the red house and my belief in its physical existence, are immanent, existent and under the scrutiny of secondary consciousness. Having admitted that a conscious act can occur at a definite level without necessarily involving any higher one, we no longer need Brentano's assumption (3), which was introduced mainly in order to block an infinite regress: if every conscious level presupposed a distinct and strictly higher one, then we should have to accept an actually infinite number of strata in every conscious act; which is clearly absurd. Brentano arrested this regress, at the secondary stage, by identifying the secondary with the tertiary level. We can however admit that every level could, but need not, be made into the existent object of a higher level. Everyday experience occurs exclusively at the primary level and issues in statements like Popper's 'Here is a glass of water', which is fallible since 'glass' and 'water' are given their usual transcendent meanings. All levels above the primary one have immanent, hence existing intentional objects; more precisely: once the higher act is performed, then its object must exist and the corresponding proposition becomes incorrigible. Of this kind are the level-o reports which describe the findings of secondary consciousness. Thus the testing of scientific theories takes place at the level of inner consciousness, through statements of the form: 'I seem to be observing a glass of water (i.e. a glass of a liquid which I take to be composed of H_2 and of O)' or, 'I seem to be seeing a red house'. Note that the modes of primary consciousness remain dependent on the theories entertained by the observer; but the role of *secondary consciousness* is exclusively descriptive of these modes. Finally, philosophical reflection is carried out on at least the tertiary level, since it examines the relationship between the primary and the inner activities.

Before accounting for the infallibility of level-o reports, let me dispel the myth that propositions can be justified only by means of other propositions. Of course, once a statement q has been deduced from another, p say, then we know that the truth of p will entail that of q. After all, by definition, valid deduction transmits truth. The above myth asserts the converse of this trivial deductive principle: it maintains in effect that all we can know is the transmission of truth, not the truth of any single proposition. Of course, if one subscribes to the correspondence theory, then Popper is right in maintaining that a truth-correspondence can never be infallibly established between a realist hypothesis and the transcendent states-of-affairs it denotes. The external world is something we do not directly apprehend; we conjecturally aim at knowing it, not by acquaintance, but through the medium of theories which cannot be effectively verified. This conclusion does not however apply to the relationship between level-o statements and the immanent states-of-affairs they describe. There are two features of autopsychological propositions which make them epistemologically privileged:

(a) They signify states-of-affairs which German-speaking philosophers appropriately call 'Erlebnisse'. As their name indicates, Erlebnisse can be 'lived through', so we have an unmediated access to them. We can thus check a level-o statement against its referent. For example, I can directly compare the sentence 'I seem (or seemed) to be seeing a red house' either with a present or with a remembered experience; but I cannot directly apprehend the state-of-affairs described by 'There exists an electromagnetic field in this room'. I am in principle unable to leap out of myself and partake of the state of the field. The privileged access to the referents of autopsychological statements does not however guarantee infallibility; for as already explained, the referent may be a remembered and hence a misremembered one. Whence the necessity of a second condition.

(b) Through inner or secondary consciousness, we can become aware of our mental acts *as we perform them*. I can observe, or rather I can sense myself observing some primary object. This eliminates or at any rate greatly reduces the possibility of my being mistaken in claiming e.g. that I now seem to be perceiving a red patch; provided of course that the claim be made simultaneously with the experience it describes. This is because, without having to rely on fallible memory, I directly compare a report about my present state of consciousness with its presumed present referent. This simultaneity also insures that all descriptive terms are given the meanings intended at the time of my uttering the report. A truth-correspondence can thus be immediately ascertained.

As remarked by Ayer, verbal mistakes are always possible. We often shout 'Right' when we ostensibly mean 'Left'. I may say 'I seem to be seeing a purple patch', where by 'purple' I now mean what was previously meant by 'red'. These are rather trivial instances of error, but there are more interesting ones. Suppose that I now see a red patch; suppose further that, yesterday, I saw a coloured patch which I correctly described as purple, but which I now wrongly remember as having had a colour identical with my present colour-sensation. However, I rightly remember having used the adjective 'purple' yesterday. As a result, I now claim to be perceiving (or rather to seem to be perceiving) a purple patch. Am I not mistaken in a non-verbal way? I do not think so: an illusion of error arises only because no proper phenomenological reduction has been performed. In 'I now seem to be perceiving a purple patch', 'purple' has been surreptitiously given a transcendent meaning not intended by the speaker. His sentence, when properly unpacked, reads: 'I (now) seem to be perceiving a colour which (now) seems to be identical to the colour which I (now) recall having seen yesterday'. This report, which is properly level-0 in both senses (a) and (b), is certainly true; it being understood that 'recall' is not to be regarded as a success word.

The two points (a) and (b) have often been conflated, except by Bretana. Initially, he even held that we are incapable of observing ourselves observing; that we should therefore always resort to fallible memory in order to reach any valid phenomenological conclusions.[23] Although Brentano later revised his views,[24] his first thesis made the *logical* distinction between (a) and (b) clear. Point (a) tells us that if we want to determine the truth-value of a sentence, not by inferring it from other propositions but by confronting it with its presumed referent, then we have no choice but to restrict such sentences to being autopsychological reports, whether these record past or present experiences. For we can directly apprehend only our own mental states. As for point

[23] F. Brentano, *Psychologie*, Vol. 1, Chapter 3 (F. Meiner, 1924).
[24] F. Brentano, *Psychologie*, Vol. 3, Chapter 1 (F. Meiner, 1968).

(b), it warns us that should we want to eliminate or greatly diminish the risk of error, then we ought further to limit ourselves to descriptions of our *present* mental states; for we know our memory to be highly fallible.

Let us now apply these last conclusions to Popper's falsificationism: in order to test scientific laws, it is no use appealing to those of their consequences which describe mind-independent events; for our access to these is exclusively theoretical. That is, we cannot break out of the vicious circle of theories, and of yet more theories, by remaining within the transcendent domain. We have a double entry only to those propositions which, on the one hand, flow from some scientific system and are, on the other hand, susceptible of being directly checked against the state-of-affairs they purportedly signify. Such statements can only be the level-o reports discussed above. Even if we feel unhappy or unsure about their incorrigibility, point (a) above clearly demonstrates that they are the only possible candidates for the status of basic statement. I have furthermore tried to show that their certainty is founded, not on any feelings of conviction on the part of some observer, but on the possibility of checking whether such propositions truly correspond to their referents. The psychologist thesis about autopsychological statements is therefore false.

8.   Unmanageable Largism?

One major difficulty remains, namely the question whether autopsychological predictions can be derived from any (appropriately extended) scientific systems. John Watkins has denied this possibility by persuasively arguing that a level-0 empirical basis would lead to unmanageable largism.[25] To my mind, Watkins's criticism is the most cogent objection brought against the phenomenological thesis. I shall nonetheless try to counter it.

It has already been admitted that the derivation of autopsychological reports greatly increases the complexity of the Duhem–Quine problem. For in deducing level-0 statements from scientific premises, not only do we have to take account of a central theory together with boundary conditions and auxiliary hypotheses; we also have to adduce extra premises about the effect which the belief in certain laws has on the experimenter's state of mind, hence on the conclusions he will draw from his observations. As part of the boundary conditions, we finally have to add assumptions about the observer's mental and physical states: if our test involves the recognition of certain colours, we may have to suppose that our experimenter is not colour-blind. Why bother with all these complications which have led philosophers like Duhem, Popper and Watkins to insist on keeping autopsychological propositions outside the proper domain of physical science? The answer, already given above, is quite simple: in case of conflict with experience, we want to have some good reason, no matter how slight, for holding our premises to be false rather than merely inconsistent with some objective statement which, by consensus, we temporarily choose to accept. As for the non-feasibility thesis, the

[25] Watkins, *Science and Scepticism*, Chapter 7.

onus of proof rests with the opponents of the phenomenological view: using Popper's transcendental method of criticism, I shall show that the actual practice of science
presupposes, if only implicitly, the phenomenological thesis.

Two simple examples will make the point. We all
agree that a colour-blind experimenter's report can be dis-
counted in situations where the perception of colour
matters. Yet a particular person's inability to distinguish
between various colours is logically irrelevant to the objective blueness, for example, of a piece of litmus paper, though
of course not to the recognition of such blueness. Hence in
chemistry, whether we are aware of it or not, the basic
statements we actually rely on are of the form 'This, which
I take to be a piece of litmus paper, looks blue' rather than
'This piece of litmus paper is blue (i.e. reflects light of a
certain frequency)'. This last objective proposition can in
fact be deduced from hypotheses which make no reference
to any conscious observer whatever. Such a derivation will
therefore not be blocked by altering assumptions about the
absence of colour-blindness; for no such assumption figures
among the premises. Only the corresponding autopsychological report requires, for its derivation, the supposition that
the observer is not colour-blind; which is precisely why
ascertaining colour-blindness could save our core theory
from refutation. Thus only the phenomenological thesis
renders the practice of science rational. Similar conclusions
apply to the chest X-ray example. The existence of some
pathological patch on the X-ray image in no way depends on
the observer's state of knowledge. Yet we should feel justified
in ignoring any experimental account given by a layman

rather than by a trained doctor. So there must be some inferential link between the observer's state of knowledge and his report. Hence the latter cannot possibly be an objective statement about some transcendent state-of-affairs; it must rather assume the form 'I notice such and such a patch which, for the following reasons, I take to be evidence for the presence in the lung of active Koch bacilli...' This statement cannot be inferred without recourse to some hypothesis about the observer's professional background; which is why a layman's report constitutes no test of any system comprising both a theory about some illness and certain assumptions about the experimenter. For the layman fails to satisfy these extra assumptions.[26]

As already explained, repeatability and intersubjective agreement should not be built into the definition of a basic statement. However, in presumed cases of refutation, we can see why the actual repetition by different people and under differing conditions of some falsifying experiment makes perfect sense, even from the phenomenological viewpoint. The main purpose of repeating a test is to render highly improbable a series of coincidences which might have falsified, in turn, various hypotheses about the non-existence of factors which are not mentioned by the theory as well as assumptions about the experimenter's mental and physical condition. Repetition enables us to hive off a number of peripheral premises and thus blame a refutation on some

---

[26] For more details, see E. G. Zahar 'John Watkins on the Empirical Basis and the Corroboration of Scientific Theories', in F. d'Agostino (ed.), *Freedom and Rationality* (Kluwer, 1989).

core theory. In this way, the Duhem–Quine problem can be made more manageable; so that the refutation of isolated hypotheses is in principle possible, if only in a fallible-conjectural way. To repeat: Popper was right to insist on repeatability and intersubjective agreement; not however because these conditions are defining characteristics of basic statements but because they constitute, circumstances permitting, a means of dealing more effectively with the Duhem–Quine problem.

Let me end by putting forward a simple logical argument in support of the feasibility of the phenomenological approach to basic statements. As long as level-0 propositions are regarded as explananda and as long as we subscribe to hypothetico-deductivism—which John Watkins and myself certainly do—then the only way we can explain such explananda is by deducing them from some set of premises; hence, by the completeness theorem, from a finite set of assumptions. Each of these assumptions either can be derived from, or else it will be adjoined to, our core hypothesis taken together with certain boundary conditions. In this way, we are bound to obtain a finite system which yields the level-0 statements as deductive consequences. Whether or not one chooses to call such a system 'scientific' is a terminological issue. The system will certainly be testable against autopsychological reports. This is why John Watkins, who so forcefully underlined the epistemological relevance of level-0 statements, ought also to accept them as constituting the empirical basis of *all* the sciences.

4 'Revolution in Permanence': Popper on Theory-Change in Science

JOHN WORRALL

Introduction

S cience, and in particular the process of *theory-change* in science, formed the major inspiration for Karl Popper's whole philosophy. Popper learned about the success of Einstein's revolutionary new theory in 1919 (the same year in which his discontent with Marxism and Freudianism reached crisis-point), and Einstein 'became a dominant influence on my thinking—in the long run perhaps the most important influence of all.' Popper explained why:

> In May, 1919, Einstein's eclipse predictions were successfully tested by two British expeditions. With these tests a new theory of gravitation and a new cosmology suddenly appeared, not just as a mere possibility, but as an improvement on Newton—a better approximation to the truth ... The general assumption of the truth of Newton's theory was of course the result of its incredible success, culminating in the discovery of the planet Neptune ... Yet in spite of all this, Einstein had managed to produce a real alternative and, it appeared, a better theory ... Like Newton himself, he predicted new effects within (and without) our solar system. And some of these predictions, when tested, had now proved successful. ('IA', p. 28)

Popper saw the development of science, through the process of change in accepted theory, as *the* exemplification of 'the critical approach'. Science is rational because all of its theories are open to empirical criticism, and because it stands ready to reject any such theory if criticism succeeds, no matter how impressively the theory had performed in the past. Having identified this approach at work in science, he went on to claim that it is the basis of human rationality both inside and outside science: what constitutes a 'mistake' may differ from field to field, but the rational method is always that of standing ready to make mistakes, and especially ready to learn from them. This simple idea then becomes a basic theme in almost every other part of Popper's philosophy.

Popper several times—especially in his later work—cited the 'simple schema'—

$$P_1 \rightarrow TT \rightarrow EE \rightarrow P_2$$

which, he claimed, characterized all rational problem-solving right across the scale from the amoeba to Einstein. In particular it is, he claimed, 'the schema for the growth of knowledge through error-elimination by way of systematic *rational criticism*' (*OK*, p. 121). Here 'P_1' refers to the initial problem situation facing the organism or scientist, 'TT' to the tentative theory proposed as a solution to the problem, 'EE' to the process of 'error-elimination' applied to the tentative theory, and P_2 to the revised problem-situation that results from this trial-and-error process.

Popper often insisted on the power of these simple ideas and clearly believed in particular that his problem-solving scheme reveals a great and simple truth. The impact of Popper's ideas suggests that, at least in some areas, this home truth really needed driving home. It is, however, difficult to think of *serious* thinkers who would challenge the scheme as it stands. Who—outside perhaps of a few extreme social constructivists—would deny that criticism and problem-solving play important roles in the development of science? Who would deny that we should learn from our mistakes and that science has managed to do just that? Who would deny that that all our theories are—to an extent at least—tentative? Certainly, the issue between Popper and rival philosophical theories of the development of science—such as those of Kuhn, Lakatos, Laudan, Shapere, van Fraassen, the Bayesians and others—is joined only when various notions are more precisely specified. What *exactly* constitutes *scientifically valuable* criticism, for example? Does producing the most valuable criticism involve holding *all* theories equally open to correction? How *exactly* is 'error' established in science? What *exactly* do we learn from our mistakes ('truer' theories or only ones that have higher empirical adequacy)? Can some theories, although always strictly speaking tentative, nonetheless become *probable* to a reasonably high degree? Are successive 'trials' informed by the successes and failures of previous ones? And, if so, exactly how? Popper, especially in his later work, insisted on interpreting various criticisms of his account of the development of science as attacks on his simple scheme and hence on 'the critical approach' in general. But, as I shall argue, some at least of these criticisms are more charitably, and more revealingly, seen as rival attempts to put some real meat on what is in truth a pretty skeletal skeleton.

The two criticisms of Popper's own attempt to fill out the details of the general scheme that I shall discuss are these. The *first* is that he basically mischaracterized the process of 'error-elimination' in science. And the *second* is that he basically mischaracterized the process by which 'tentative theories' are proposed. Put baldly: Popper's view was that science is entirely based on the method of 'trial-and-error', 'conjecture and refutation', and yet—so these criticisms allege—he seriously misidentified the nature *both* of the process of identifying error in science *and* of the process of theory-production or 'conjecture'. Both criticisms, especially that concerning error-elimination, have been developed and quite extensively discussed before. I hope however that I shall add something new to them.

1. Refutations: Popper *versus* Kuhn, Lakatos *et al.*

It is easy to get the impression when reading Popper (or, rather, it is difficult to avoid the impression) that the basic picture being presented is a very straightforward version of the trial-and-error schema. Tentative theories are put forward in response to problems; these theories are scientific only if empirically testable—that is, only if they have deductive consequences whose truth value can be agreed on in the light of experiment or observation; those theories are tested; some may, if we are lucky, survive for a while the severest tests we can subject them to—these theories are *temporarily* 'accepted'; but the process of testing must always continue and if a hitherto accepted theory eventually fails a test, then it is rejected and a new tentative theory sought. The chief vehicle of scientific progress, on this straightforward view, is the direct empirical refutation of theories.

Criticism of Popper's emphasis on experimental falsification came to a head as a result of the impact of Thomas Kuhn's views of science. There is, however, nothing of real relevance to this particular issue in *The Structure of Scientific Revolutions* that was not raised already in Duhem's *The Aim and Structure of Physical Theory*.[1] Indeed many of the Kuhnian theses that have created such a stir in philosophy of science seem at root to be (often rather less clear) restatements of Duhemian positions. Consider, for example, Kuhn's famous claims about 'elderly hold-outs', claims that, as we shall see, take us straight to the heart of the falsifiability issue.

According to Kuhn, if we look back at any case of a change in fundamental theory in science, we shall always find eminent scientists who resisted the switch to the new paradigm long after most of their colleagues had shifted. Famous examples of such hold-outs include David Brewster, who continued to believe in the corpuscular theory of light long after Fresnel's wave theory was developed, and Joseph

---

[1] T. S. Kuhn, *The Structure of Scientific Revolutions* (first edition, 1962; second edition 1970, Chicago: Chicago University Press). P. Duhem, *La Theorie physique. Son objet, sa structure* (Paris, 1906); English translation *The Aim and Structure of Physical Theory* (Princeton: Princeton University Press, 1956).

Priestly, who persevered in defending the phlogiston theory against Lavoisier's oxygen theory. These hold-outs are often (though not invariably) elderly scientists who have made significant contributions to the older paradigm.

Kuhn is of course right that important hold-outs existed; he may or may not be right that there are *always* significant hold-outs in any scientific revolution. But the challenging Kuhnian claim was not descriptive, but instead the *normative* claim that these elderly hold-outs were no less justified than their revolutionary contemporaries: not only did they, as a matter of fact, stick to the older paradigm, they were moreover, if not exactly right, then at least *not wrong* to do so. On Kuhn's view, 'neither proof nor error is at issue' in these cases, there being 'always some good reasons for every possible choice'—that is, *both* for switching to the revolutionary new paradigm *and* for sticking to the old. Hence the hold-outs cannot, on Kuhn's view, be condemned as 'illogical or unscientific'. But neither of course can those who switch to the new paradigm be so condemned.

Why does Kuhn think the resisters no less rational than their more mobile colleagues? His full account of theory change is complex, but he is quite clear about the basic reason why resistance of the new theory fails to be irrational:

> The source of resistance is the assurance that the older paradigm will ultimately solve all its problems, that nature can be shoved into the box the paradigm supplies.[2]

[2] Kuhn, *The Structure of Scientific Revolutions*, pp. 151–152.

The fundamental point here is the one already made by Duhem, namely, that the sorts of assertions that tend to be called 'single', 'isolated' theories have in fact no empirical consequences of their own. A scientist may speak of testing Newton's theory of gravitation, say, by observing planetary positions. But when that test is subjected to a full deductive analysis, it is readily seen that a range of other assumptions are in fact involved—amongst others, assumptions about the number and masses of the other bodies in the solar system, about the non-existence (or neglibility) of any forces other than gravitational forces, about how the telescope works and about the extent to which light is refracted in passing into the earth's atmosphere. All these assumptions are needed if a genuinely observational consequence is to be deduced. Thus the smallest unit with anything like directly testable empirical consequences is a *group* of theories (or a *theoretical system*) based on a central theory, but including more specific assumptions and auxiliary and instrumental assumptions.

Often—as, for example, in the case of the corpuscular theory of light (a case that Duhem himself considered in detail)—the situation is further complicated: the 'central theory' itself breaks down into a 'core' and a set of more specific assumptions. Thus 'the' corpuscular theory of light consists of the basic assumption that light consists of *some sort* of material particles; to which scientists then need to add more specific assumptions about the particles (for instance about what differentiates those producing blue light from those producing red light), and about what particular forces act on those particles in particular circumstances (for instance in passing from one optical medium into another).

Duhem's point about the real deductive structure of observational and experimental tests implies, of course, that if a test outcome is negative, if the experiment or observation contradicts the predicted consequence, then, even assuming that we know the (negative) test outcome for sure, the only theoretical unit that we can strictly infer is false is the *whole set* of assumptions needed to derive the experimental prediction. That is, all that we know (directly) from such a 'refutation' is that at least one assumption from within this set is false, we do not know just from the negative result which *specific* assumption is false. In particular, we cannot of course infer that it is the 'central' theory.

It is easy to point to historical cases in which scientists retained a central theory despite the experimental refutation of the theoretical system based on it by rejecting instead an auxiliary assumption. And, importantly, it is easy to point to such cases in which *the scientists concerned seem obviously justified in doing so*. One especially famous example concerns the discovery of the planet Neptune. The predictions about the orbit of the planet Uranus made on the basis of Newtonian theory turned out to be wrong. But instead of regarding this as refuting that theory, Adams and Leverrier independently conjectured that there is a further and hitherto undiscovered trans-Uranian planet, and that once its gravitational effect is taken into account the correct predictions about Uranus's orbit will follow from the theory. Adams and Leverrier were in effect pointing out that a prediction about a particular planet's position cannot be deduced just from Newton's theory; instead, further assumptions are required—in particular one about the other gravitating bodies that are affecting the planet concerned. And they went on to suggest that the best way to deal with the refutation of the initial theoretical system by the observations of Uranus was, *not* by rejecting the central Newtonian theory, but by rejecting the initial auxiliary about the number of other bodies affecting Uranus. Roughly speaking, they 'worked backwards'—assuming the truth of Newton's theory—to discover the simplest assumption that would give the right empirical results, and this turned out to be that there is a further planet beyond Uranus that astronomers had not yet noticed. This claim turned out to be dramatically confirmed.

In cases like that of 'the' corpuscular theory there is a further choice available to the scientist in the event of a clash between his overall theoretical system and evidence. This is an option that might loosely be described as 'modifying' the central theory rather than rejecting it entirely. In these cases, the central theory (as just indicated) itself has a 'core claim' (in the case of the corpuscular theory of light, the claim that light consists of *some sort* of material particles) alongside more specific assumptions (for example, assigning masses and velocities to the particles producing different kinds of light, making particular assumptions about the forces that affect these particles, and so on). A proponent of the corpuscular theory of light might find that, when she makes particular assumptions about the forces that operate on the light particles entering a transparent medium like glass, and adds plausible auxiliaries about her instruments, her overall system is refuted by observation of the amount of refraction the light actually undergoes. Seeing no way to replace the auxiliary assumptions she is making, she may decide nonetheless that it was her specific assumption about the forces operating on the particles at the interface that was wrong rather than the general assumption that light consists of material particles of some sort subject to some sort of forces. She will then produce a new theoretical system with the same auxiliaries as before and with the same core idea of the corpuscular theory, but with different specific assumptions about the particles and the forces on them.

The upshot of Duhem's analysis, then, is that when the deductive structure of a test in science is fully analysed, a whole group of assumptions is needed to derive the observational result—none of the assumptions alone being strong enough to entail any such observation. It follows just by deductive logic that, for any single specified theoretical assumption T, there must be for any set of observation statements a group of assumptions including T that entails all the observation statements in the set. Thus there are no crucial experiments: no result or set of results ever *forces* a scientist logically to give up any single theory. Duhem pointed out for example that any of the famous alleged crucial experiments against the corpuscular theory of light could have been accommodated within the corpuscular theory 'had scientists attached any value to the task'.[3]

---

[3] Duhem, *The Aim and Structure of Physical Theory*, p. 187.

Kuhn's discussion of hold-outs is, in large part, just a corollary of this Duhemian analysis. Kuhn simply adds to Duhem the historical claim that there are always (or usually) *some* scientists who 'attached some value to the task' of accommodating the allegedly crucial counterevidence within the older framework. But does Kuhn's striking claim that these hold-outs were 'neither illogical nor unscientific' also follow from the Duhemian analysis?

Kuhn is surely right that the hold-outs cannot be faulted as 'illogical': *if* deductive logic is the only constraint, then the hold-outs' insistence that the evidence regarded as crucial by the revolutionaries *can* be 'shoved into' the box provided by their favoured older paradigm not only fails to be demonstrably false, it is demonstrably true. Given that the paradigm-constituting central theory T, has no directly checkable observation sentences as deductive consequences, it follows that for any set of such observation sentences there must always exist a consistent theoretical system that entails T and also all the observation sentences in the given set.

But what of the claim that these hold-outs also fail to be 'unscientific'? Kuhn seems implicitly to assume (and the many sociologists of science influenced by him quite explicitly assume) that it follows from the fact that accommodation of the allegedly crucial result within the older system is always logically possible, that reason is powerless to judge against the hold-outs. But this, in effect, identifies scientific rationality with deductive rationality. An alternative conclusion (and surely the correct one) is that there are *further* articulable principles of scientific rationality which differentiate those cases where it is reasonable to defend a theoretical framework in the way that Duhem indicates is always logically possible from those cases where it is not reasonable.

Adams and Leverrier showed how to 'hold onto' Newtonian physics by attributing the clash between its predictions about Uranus's orbit and observations of that orbit to the omission of the gravitational influence of a hitherto unrecognized planet. Gosse, as is equally well known, showed how to 'hold onto' the theory that God created the universe with all its present 'kinds' in 4004 B.C. by attributing the apparent evidence of now extinct species to God's decision to include in His creation things that look remarkably like bones of organisms of earlier species or remarkably like the imprints in the rocks of the skeletons of such organisms. There is surely a crucial difference between these two cases—a difference that has much to do with independent testability (the Adams and Leverrier switch makes new testable predictions—for example about the orbit of the new planet, while the 'Gosse dodge' is designed precisely to accommodate initially threatening data while permitting no further tests), and perhaps something to do with judgments of the relative 'plausibilities' of different possible auxiliary assumptions in the light of background knowledge. The main task that Kuhn's analysis sets for the holder of the view that radical theory-change in science is a rational process is—so it seems to me—precisely that of first articulating the principles that underwrite the distinction between these two types of case, and then showing that those principles are themselves rationally defensible. It is exactly this task that Lakatos, Laudan and subsequently many others have undertaken with varying degrees of success.

Duhem's points have often been emphasized in recent discussions and what I've said might seem like stressing the by now obvious. Duhem's simple and vitally important message is, however, still often misconstrued. Two respects in which my account differs from some others in the literature should be given special emphasis. *First*, what is true about theory-testing is that when the deductive structure is properly analysed the set of necessary premises is quite *large—larger* perhaps than one might at first expect. What is *not* true (at least not in any interesting sense) is that 'the whole of our knowledge' (whatever that may be!) is involved in the attempt to test any part of it. The slip from 'we need lots of assumptions to get consequences that are really directly checkable empirically' to 'there is no end to the assumptions we need' is, I think, just sloppy: all deductions are finite.[4]

---

[4] Quine seems to make this slip. This seems to be partly based on a failure consistently to distinguish between 'Indefinitely many assumptions are needed if any observational statement is to be derived' (false) and 'Although "only" finitely many assumptions are needed and so at least one of that finite set must be rejected if the observational consequence proves false, there is no *a priori* limit on the assumptions that might in turn be affected by that initial rejection' (true, but unsurprising). It also seems partly based on a flirtation with the idea that not even *deductive logic* can be taken as fixed here. But if not even a *core* of logic is taken as constituting the framework of the discussion, then it is not clear that any sense can be made of *any* assertion about testing.

The *second* point on which I disagree with some other Duhem-inspired analyses concerns the allegedly inevitable fallibility of basic statements. Suppose Newton's theory of gravitation, for example, is being tested by observations of the position of some planet, say Neptune, at some given time. One formalization of the deductive structure of the test will involve Newton's four laws as the only explicit general premises, plus 'initial conditions' about Neptune, along with positions and masses of the other bodies in the solar system at time t, and finally a 'closure' assumption—of the form 'only gravitational effects have any non-negligible effects and the only non-negligible gravitational effects are those produced by the other massive bodies in our solar system: the sun and the planets aside from Neptune itself.'[5] A conclusion will then be drawn (actually only with the help of mathematical approximations—another story that I shan't go into here) about the position of Neptune at time $t+\Delta t$. No mention in this formalization of optics, atmospheric refraction or the rest. But of course the acceptance of the 'initial conditions' concerning planetary positions at t, as well as that of the test prediction about the position of Neptune at $t+\Delta t$, depend implicitly on auxiliary theories about optics. Scientists can only be construed as 'observing' planetary positions by pointing telescopes at the sky if a range of background theories is taken for granted.

[5] An alternative formulation would simply involve an 'initial condition' about the *total* force acting on Neptune at t. But the more complex assumptions about the planets as well as the 'closure' assumption would, of course, simply be hidden in such an initial condition.

There are no rules about how to formalize informal deductions, and so long as the auxiliary assumptions concerned are regarded as uncontroversial it will be natural to leave them implicit. This formalization, however, involves initial conditions and a test prediction that are theory-laden, not just in the trivial sense that all statements about the objective world are bound to be (the assertion that Descartes's tedious demon does not exist is of course a theory), but in a sense that involves the serious possibility of later correction. If the test is formalized in this way, then there appear to be two options in the case of an inconsistency between 'observation' and theory: reject the theory (still in fact a theoretical *system*, though a comparatively slim one) or reject the test result—where the latter means asserting that *either* the initial conditions in fact failed to hold *or* the apparent result was 'wrong'.

Some episodes from history of science are naturally told as ones in which the second option was adopted: some observational claim was 'corrected' in the light of theory. A celebrated example concerns Flamsteed and Newton. Again speaking very roughly: Newton wrote to Flamsteed, the first Astronomer Royal, to ask him to check some of his theory's predictions about planetary positions; Flamsteed wrote back telling Newton that the predictions were incorrect; Newton replied that his theory is correct, it was Flamsteed's observations that were wrong and if he cared to 'recalculate' them, using the formula for atmospheric refraction that he, Newton, supplied, then he, Flamsteed, would find that his observations really confirmed the theory's predictions. Since it was Newton's view that prevailed, this looks like a classic case of theory overriding observation and hence not just of the in principle theory-ladenness of observation but of the real in practice corrigibility of observation statements.

There is nothing wrong with this way of telling the story so long as it is realized—as it seldom is—that it is equivalent to the following, perhaps less exciting, but more revealing account. The clue to the second formulation of the test is Newton's suggestion that Flamsteed should *recalculate* his data. This implies, of course, that there was some 'crude data'—basically records of the angles of inclination of certain telescopes at certain times (that is, when certain clocks showed particular readings)—which were never questioned in this episode. In order, however, to 'calculate' planetary positions from these 'crude' data, Flamsteed had to make various assumptions of a low-level, but nonetheless clearly theoretical, kind. These assumptions included one about the amount of refraction that light undergoes in passing into the earth's atmosphere. If these assumptions are teased out and added to Newton's theory plus the original auxiliaries, then this creates a still larger theoretical system that, unlike the original, has deductive consequences at the crude data level. Using this more extensive articulation of the test, this episode is restored to one in which unchallenged data clashed with a theoretical system and in which the dispute was simply over which of the assumptions making up that system should be rejected. Flamsteed was suggesting that it should be Newton's theory and Newton that it should be Flamsteed's (theoretical) assumption about atmospheric refraction.

It is sometimes supposed that there are two separate reasons why empirical refutations of theories can never be conclusive: the Duhem problem *and* the inevitable fallibility of basic statements. Less confusion results, I claim, if we recognize only one problem—a 'big Duhem problem'. In any interesting historical case, there is always a level of 'data' low enough so that all sides in a dispute agree on the data. It's just that a very large number of assumptions need to be articulated and included in the theoretical system under test, if sentences at that level are really to be deduced from that system; and this means of course that, if the theoretical system turns out to be inconsistent with the (crude, unchallenged) data, there are a large number of options for replacing some part of the system to restore consistency.

In sum, the lesson that ought to have been derived from Duhem's analysis is then the following. It is always large theoretical systems that clash with empirical results; but if the system is made large enough, *every* such clash can be represented as one in which the empirical result at issue was, if not entirely unquestion*able*, then certainly never questioned. The chief methodological problem bequeathed by the analysis is simply that of formulating rules for ranking different modifications of such systems in the light of initially refuting evidence. Why did Adam's and Leverrier's 'modification' of the Newtonian system strengthen the basic Newtonian gravitational theory, while the 'Gosse dodge' does nothing to strengthen the empirical support for the basic idea of creationism?

The consequence of Duhem's analysis for the idea that science is characterized by the falsifiability of its claims seems quite straightforward and yet has often been obscured. Of course we can use terms as we wish, but it seems prudent in view of the above to say that there are indeed falsifications in science but that what gets *falsified* or *refuted* are large theoretical systems and never 'single' scientific theories. (Lakatos in particular was responsible for large amounts of terminological confusion here: talking, in one breath, of theories ('single', 'isolated' theories) as 'irrefutable' and, in the next, of scientists 'saving' such theories from refutation.) Duhem's analysis does *not* imply that there are no arguments for the falsity of 'single' theories provided by science. It implies only that such arguments do not consist of the deduction of a false observational sentence from that single theory, nor even of a deduction of such a false observation sentence from the theory *plus other relatively uncontentious assertions*. Instead the argument for the falsity of theories such as the claim that light beams are made up of material particles of some sort consists of two parts. First, a demonstration (usually based on a long sequence of reactions to refutations of theoretical systems built around the same core) that what other assumptions are needed in order to render that theory consistent with data is (a) implausible and (b) (crucially) entirely lacking in *independent* empirical support; and *secondly* of the production of a rival core theory, inconsistent with the first, and a demonstration that this second core theory can be incorporated into a system which enjoys *independent* empirical support and whose auxiliaries are at any rate much *more* plausible. So, for example, there is surely no doubt that the idea that light consists of material particles is false. The argument for its falsity, however, depends on no empirical refutation but (a) on the fact that various phenomena (for example, those of diffraction) could only be incorporated into particulate theoretical systems in an entirely ad hoc, non-independently testable fashion and (b) on the fact that those phenomena 'fell out naturally' from theoretical systems built around the rival wave theory.[6]

Kuhn's elderly hold-outs, then, did not perversely fail to see an experimental refutation of their favoured central theory. They were not wrong in that precise sense, but that does not mean that there was no objective argument based on experiment which told against their position. The argument that their favoured central theory eventually came to be regarded as false *for good* (*empirical*) *reasons* does not, however, consist of an empirical refutation (not even 'in hindsight'), but of the demonstration that systems built around that central theory 'degenerated' while systems built around a rival central theory scored impressive independent empirical success.

2. Falsifiability and 'conventionalist stratagems': was Popper a Duhemian about 'refutations' all along?

The Duhemian point that underlies this line of criticism of the idea of direct falsifiability of 'isolated' scientific theories

---

[6] For further details see my 'Falsification, Rationality and the Duhem Problem: Grünbaum vs Bayes', in J. Earman, A. I. Janis, G. J. Massey and N. Rescher (eds), *Philosophical Problems of the Internal and External Worlds* (Pittsburgh and Konstanz: University of Pittsburgh Press, 1993).

seems both simple and undeniable. And indeed one aspect of Popper's reaction to Kuhn and to Lakatos was that, far from being a criticism of his position, he had emphasized it all himself long ago. Popper pointed to his explicit assertion made, not in response to Kuhn, but already in the original *Logik der Forschung*, that 'no conclusive disproof of a theory can ever be produced' (*LSD*, p. 50), and to his explicit acknowledgment that one reason for this was the ever present possibility of 'evading' an attempted refutation by using 'conventionalist stratagems' (where one type of conventionalist stratagem involves 'introduc[ing] *ad hoc* hypotheses').[7]

Popper *does* seem to have made the mistake—both in 1934 and later—of thinking that auxiliary assumptions are only ever *introduced* in order to 'save' a theory. In fact Duhem's point was of course that, whether we are aware of it or not, such auxiliary assumptions are *always* involved in empirical tests. For example, in discussing the Lorentz–Fitzgerald contraction hypothesis, Popper talks as if an assumption about the length of the interferometer arms was *introduced* as a result of the null-outcome of the Michelson–Morley experiment—the assumption, that is, that the length varies depending on the velocity of the arm through the ether. But clearly classical physics cannot predict any outcome of the experiment without making *some* assumption about the length of the arms: before the

[7] The 'other' reason given by Popper for the inevitable inconclusiveness of empirical refutations is the alleged inevitable fallibility of basic statements. In fact, as indicated *above*, this is best treated as another, rather confusing way, of putting the same Duhemian point.

null-result, however, this was the 'natural' assumption that the arms were always the same length (since this is what 'solid rod' congruence measurements revealed). This slip is not as minor as might at first appear: it led Popper to make the further, related mistake of supposing that good scientific practice demands that 'conventionalist stratagems' are always to be avoided. He wrote (*LSD*, p. 82):

> We must decide that if our system is threatened we will never [sic] save it by any kind of *conventionalist stratagem*. Thus we will guard against exploiting the ever open possibility ... of ... attaining for any chosen ... system what is called its 'correspondence with reality'.

But this itself might only betoken a minor confusion, for he immediately added the important remark (ibid.):

> As regards *auxiliary hypotheses* we propose to lay down the rule that only those are acceptable whose introduction does not diminish the degree of falsifiability or testability of the system in question, but, on the contrary, increases it.

Popper characterized one form of conventionalist stratagem as involving the 'introduction' (in fact, modification) of auxiliaries ad hoc—that is, in a way simply designed to solve the problem posed by some refutation of the earlier theoretical system. It seems to follow then from this rule about auxiliary assumptions—accentuating the positive—that *if* a 'conventionalist stratagem' does not diminish, but in fact *increases* the degree of testability of the system, then it counts as a scientifically acceptable move. This, however, clearly contradicts his claim, made only a few sentences earlier, that in order to be scientific we must *always* eschew conventionalist stratagems. If this contradiction is resolved by taking the remark about independent testability as definitive, then it would seem to point us in the (right) direction—towards the acceptance that (relatively common) experimental falsifications falsify only large theoretical systems, and towards the acceptance that the really major experimental results are those that *confirm* the excess predictions of one such system compared to available rivals.

Popper's remarks from 1934 supply material, then, for something like the following reply to his 1960s critics: 'Kuhn and Lakatos are pointing to a slight inconsistency in the presentation of my original position, but once this is resolved in a fairly obvious way then my position in fact anticipates the point they make in alleged criticism of it. We are all agreeing on a Duhemian analysis of testing. *Of course* I agree that "single" theories are never testable in isolation. But theories like Newton's have been properly tested as parts of theoretical systems which involve independently confirmed auxiliary assumptions, while theories like Freud's have never been incorporated within genuinely testable systems, but only—at best—within systems in which already known experimental results have been accommodated after the event. When I said that Newton's theory is testable and Freud's isn't, this was simply shorthand for this rather more complicated claim.'

Although it runs counter to a lot of the straightforward falsificationist rhetoric in Popper's various writings, this would, I believe, indeed be Popper's response in a rational reconstruction of history. How far does the reconstruction caricature real history? I have been unable to find a satisfactory answer. Many of Popper's remarks suggest that he genuinely felt that Lakatos in particular was deliberately setting up a strawpopper only to use the real Popper's ideas to knock him down; and yet Popper's reactions in the Schilpp volume to both Lakatos and Putnam (who makes much the same point about falsifiability) seem monumentally to miss the point.

Popper's 'Replies' ('R.C.') contain a number of suggestions about the issue of falsifiability that do not obviously cohere with one another. There is some consideration of the old favourite 'all swans are white'. (Such examples are, of course, beside the point since Duhem's analysis applies to proper scientific theories, not simple observational generalizations, which clearly *do* have observational consequences 'in isolation'.) There is also some suggestion that Popper was going to show that Newton's theory unlike Freud's *really is* refutable, that Newton's theory can be brought into direct conflict with observational potential falsifiers without needing 'auxiliary assumptions' (which Popper insisted he had taken account of all along under the name of 'initial conditions'). In substantiation of this Popper remarked 'If the force of gravity were to become a repulsive force Putnam would soon notice the difference...' ('R.C.' p. 998). But of course the issue is not about a possible *change* in the laws of nature (even if that notion can be made sense of); and 'the force of gravity is repulsive' is clearly another theory not an observation statement. (Any theory, Freud's for example, is of course falsifiable if we allow any claim, no matter how theoretical, to count as a potential falsifier—for example (?), 'no one was ever affected psychologically by any real or imagined sexual trauma in childhood.') Popper elsewhere made more sense of this kind of response by claiming that, for example, the observational claim 'Mars (say) moved in a square orbit' is a potential falsifier of Newton's theory. But of course—Duhem's old point—that claim *is* consistent with Newton's theory taken in isolation. This is not to deny that, had such an observation been made, Newton's theory of gravitation would have been rejected. But this would *not* be on account of a direct refutation unmediated by auxiliary assumptions. Instead it would be because it seems clear in advance that any auxiliary assumptions that would produce a Newtonian theoretical system that implied the square orbit would have been massively implausible on other grounds. (Remember, no one is denying that there are circumstances under which the only reasonable thing to do is to give up some single scientific theory—the dispute is only over the *reason* for this.)

In the end it is quite unclear what position Popper held in the Schilpp volume 'Replies' ('R.C.'). On p. 998 he asserted—apparently unambiguously—that, while auxiliary assumptions might be needed for certain sophisticated tests, Newton's theory permits also certain 'crude' tests where such further assumptions are unnecessary (and so 'Newton's theory can be refuted without the help of initial conditions (i.e. auxiliary assumptions)'). But then in replying to Lakatos on p. 1004, although Popper continued to insist on the falsifiability of Newton's theory in contrast to Freud's, *this was only* '[d]isregarding the possibility of immunizing strategies'. Since this possibility and the consequent need to distinguish progressive and degenerating shifts in theoretical systems was the whole issue, I find it hard to know what sense to make of a discussion that explicitly 'disregards' them. (This point has also been made by John Watkins.[8]) Duhem's analysis shows how to avoid any loose talk about 'falsifying hypotheses', 'fallible falsifications' and the like. A genuinely falsifiable theoretical system cannot be 'immunized' against a falsification; and as for unfalsiable single theories, there is no need to immunize them against non-existent falsifications.

I gratefully leave further exegetical investigation to others and settle for the following qualified conclusion: so far as the falsifiability aspect of the 'Popper–Kuhn' or 'Popper–Lakatos debate' goes, *either* there was never really anything at issue or Popper lost. There is, of course, 'error-elimination' in science but only, *directly*, of large theoretical systems. The way in which components of such systems come to be regarded as errors (in particular the way in which the 'central' components come to be regarded as false) is an altogether more complex process than simple empirical refutation. If some remarks in Popper suggest that he had insight into this process, others seems to suggest he had little, and in any event both Kuhn and Lakatos were altogether more successful in describing the details.

3. Conjectures: Popper versus Kuhn, Lakatos *et al.*

As in the case of theory-refutation, there is a simple and straightforward account of theory-*production* that it is difficult

---

[8] John Watkins, *Science and Scepticism* (Princeton: Princeton University Press, 1984).

to avoid taking away from Popper's writings. This account sees the trials in the trial and error method of science as 'random', or better 'blind'—uninformed by, not deliberately structured to meet, the epistemic environment into which they are to be launched. Just as mutations in Darwinian theory are not environment-directed, nor pre-designed to solve some range of ecological problems or to fit some existing 'ecological niche', so scientific theories, on the view Popper seems to defend, are not predesigned to meet existing epistemic problems: they are *generated* in a way that is blind to existing epistemic needs, although, once generated, they are subjected to rigorous selective pressure. As Popper put it (*OK*, p. 144):

> The growth of knowledge—or the learning process—is not ... a cumulative process but one of error-elimination. It is Darwinian selection, rather than Lamarckian instruction.

As in the case of refutations, this simple view is quite seriously and straightforwardly incorrect. Again the first and most important thing is to see clearly just how wrong the view is. The purely exegetical issue of how far, or how consistently, Popper really held the view is of secondary importance and is accordingly held over to the next section.

Kuhn's criticism of the idea that scientific theories are refutable is of course a relatively minor part of the account of science developed in *The Structure of Scientific Revolutions*. His fundamental idea is that of 'normal science' practised under the aegis of a 'paradigm', where a paradigm involves not just substantive assertions about the world, but also heuristics or 'puzzle-solving techniques' (and some other, perhaps less reputable, things besides). Puzzle-solving techniques (and the idea of emulating 'exemplars') guide the construction of specific theories and in particular guide the reaction of normal scientists to experimental 'anomalies'. Although there is much in Kuhn's elaboration of this idea with which to disagree, the fundamental idea itself is surely correct and important. At any stage of the development of science, there are ideas about how to construct theories and how to modify theories should experimental difficulties arise, ideas that can, with effort, be sharply articulated and are as much a part of 'objective knowledge' as theories themselves. Alternative elaborations of this basic idea were being suggested around the same time by Hanson, Hesse, Post and others; and Lakatos's idea of a research programme, characterized in part by its 'positive heuristic' was an attempt—not an entirely successful one—to characterize this aspect of science more precisely and in more detail. There was also a largely independent development of the idea of rational heuristics within the Artificial Intelligence literature, beginning with Simon and Newell in 1958. More recently several philosophers of science have developed more detailed views along the same lines; while the idea of producing AI programs that will generate scientific theories, though still relatively in its infancy, has already begun to produce interesting results.[9]

---

[9] A quite comprehensive review of both the philosophical and AI literatures on rational heuristic can be found in chapter 2 of Ken Schaffner's recent *Discovery and Explanation in Biology and Medicine* (Chicago: Chicago University Press, 1993). See also Elie Zahar's *Einstein's Revolution: A Study in Heuristic* (La Salle: Open Court, 1987).

The fact is that scientists don't simply guess their theories; they don't make 'bold' Popperian conjectures. Instead they arrive at their theories in a way that, while it no doubt involves intuition and creativity, can nonetheless be reconstructed as a systematic, and logical argument based on previous successes in science and parts of 'background knowledge' taken as premises. Kuhn's own attempt to flesh out this suggestion was not, I believe, especially successful; and although others, notably Elie Zahar, have in the meanwhile done better, a full analysis is still some distance away. However, even in the absence of a general analysis of theory-construction, it is not difficult to show, by examining the *details of particular historical episodes* that such an analysis must exist.

Two sorts of situation recur time and again in the history of science. In one sort of case a general theory becomes accepted—in part because some specific theory based on it scores impressive empirical success; and it then proves fruitful to take (of course for the time being) that general theory as a premise and try to develop further specific theories based on it to account for further, related ranges of phenomena. It seems to be a fact about the history of physics at least that this happens often *and works*. The second related sort of case is where some general theory has been accepted, some specific theories based on it are successful, but the latest such specific theory developed out of that general theory, though initially empirically successful, then runs into empirical anomalies; scientists have then continued to assume the general theory as a premise and tried to use the anomalies to argue a different specific theory on the same general lines. Again this has often proved successful.

These two possibilities are readily illustrated in the relatively straightforward case of the classical wave theory of light. The general theory that light consists of vibrations transmitted through an all-pervading mechanical medium was shown by Fresnel's treatment of diffraction in 1818 to be highly empirically successful. This treatment amounts of course to a specific theory based on the general wave idea. As is well known, Fresnel's theory of diffraction predicted, for example, that the centre of the 'shadow' of a small opaque disc held in light diverging from a single slit would be bright—indeed that the very centre would be just as intensely illuminated as if no disc were held in the light. And this prediction had been verified by Arago.[10] When Fresnel and later others came to deal, then, with further optical phenomena, such as the transmission of light through birefringent crystals, it was natural to use this general assumption of light as waves in a medium as a basis. The heuristic guidance given to scientists in this way is much stronger than might be supposed and, by dint of detailed analysis of particular historical cases, can be much more clearly articulated than was managed by either Kuhn or Lakatos or others who discerned this important point.

So, for example, when Fresnel came to develop his account of the transmission of light through crystals, he took it that he was looking for an account of the form of the wave

---

[10] For the real story of this historical episode see my 'Fresnel, Poisson and the White Spot: the role of successful predictions in the acceptance of scientific theories', in D. Gooding, T. Pinch and S. Schaffer (eds), *The Uses of Experiment* (Cambridge University Press, 1989).

surface within such crystals. He took it that the ether within the crystal is a mechanical medium—that is, that when one of the ether's parts is disturbed from its equilibrium position it is subjected to an elastic restoring force. Observation had shown that there are three types of transparent media: 'ordinary' unirefringent ones (ones in which only one refracted beam is created), and two different classes of birefringent media (so-called uniaxial and biaxial crystals). Previous studies of elastic media in general had established —it was part of 'background knowledge'—that the elastic restoring force acting on a part of the medium drawn away from equilibrium generally depends on the direction of the disturbance. Such direction-dependence could be expressed mathematically in terms of the coefficients of elasticity along three arbitrarily chosen mutually orthogonal axes through the medium. Fresnel was hence led to the theory that the three types of transparent medium are ones in which (a) all three coefficients of elasticity are different, (b) two coefficients are the same and the third different, and (c) all three coefficients are the same. Case (c) is the isotropic case of unirefringent media; case (b) is uniaxial birefringents crystals; case (a) biaxial birefringent crystals.

In the second type of case that I mentioned, the latest specific theory, developed out of some general idea, though initially successful, then runs into experimental anomalies, but a new specific theory is then looked for *based on the same general theory* together with *the initially anomalous data*. This can again be precisely illustrated in the case of Fresnel. Background knowledge in the form of accepted theories in the mechanics of 'continuous media' entailed that two types of waves can be transmitted through such media: pressure waves produced by the medium's resistance to compression and waves produced by the medium's resistance to shear (if any). The former are longitudinal—that is, the vibrations of the parts of the medium constituting the wave are performed in the same direction as that of the overall transmission of the disturbance through the medium. The latter are transverse—the vibrations occur at right angles to the transmission of the whole waveform. Fluids transmit longitudinal pressure waves; only solids exhibit resistance to shear and hence only solids can transmit transverse waves (along, in general, with longitudinal ones). Since the light-carrying ether has to allow the planets to move freely through it (background knowledge again entailing that, to within observational error, the planets' motions are accounted for simply by gravitational effects), Fresnel (like all his predecessors) naturally took it that the ether is a (very highly attenuated) *fluid* and hence that light waves are longitudinal. However this specific assumption was refuted (as, of course, part of a theoretical system involving further assumptions) by the results of Fresnel and Arago's modified version of the famous two slit experiment.

Fresnel and Arago found that when plates of polarizing material are placed over the two slits in this experiment in such a way that the beams emanating from the two slits are oppositely polarized, the previously visible interference pattern is destroyed. Oppositely polarized light beams fail to produce interference fringes and in particular fail to interfere destructively for any values of the path-difference. But near the centre of the observation screen the two beams are nearly parallel, and so the vibrations making them up would also be parallel, if the waves were longitudinal. This in turn implies that they could not fail to interfere destructively at those places of the observation screen corresponding to path-differences of odd numbers of half-wavelengths (assuming, of course, that the general view is correct, that is, that the beams *do* consist of some sort of waves in a mechanical medium).

One possibility in view of this difficulty would be to give up the general theory, but that general theory had been impressively successful elsewhere, and there was not the slightest indication of how to look for a different general account that could do anywhere near as well. Hence Fresnel's attitude was essentially that, given that the general theory had to be correct, the only serious question was what this experiment was telling us about the vibrations that were known to exist. And the answer was clearly that the waves (or at any rate that part of the waves responsible in general for interference effects) cannot be longitudinal. Background knowledge tells us that the other possibility is transverse waves and so Fresnel inferred (and, given his general assumptions and background knowledge, it can be reconstructed as a genuine *deduction*) that the light vibrations are orthogonal to the direction of light-propagation and, in the case of oppositely polarized rays, orthogonal to one another.

Even this rather sketchy account shows that there is nothing in this process that even remotely resembles the 'bold conjectures', the constant ferment of ideas, the 'revolution in permanence' so beloved by Popper. Refutation plays an important part in this second case but Fresnel's response to the refutation in his initial theory can hardly be described as producing a bold conjecture that then just happened to survive criticism for a while. Instead Fresnel arrived at his new, transverse theory by a systematic, deductive process, using background knowledge and the results that had refuted his earlier theoretical system. Scientists use claims that they regard as relatively well-entrenched in order to deduce specific theories from experimental discoveries. Of course if *any* step from one detailed theory to another inconsistent with it counts as a revolution (as some of Popper's remarks about 'little' revolutions suggest), then science no doubt *would* count as 'revolution in permanence', but this phrase gives all the wrong signals. The changes at issue are produced by *depending* on background assumptions of various degrees of generality. These background assumptions are used (however temporarily) as 'givens' or premises. It is the relative fixity of its theories, not their constant revolutionary change that accounts for the success of science. This is not of course, as Popper suggested in reply to Kuhn, a case of unmotivated dogmatic attachment to theories. There is in any event nothing immutable about the premised background assumptions (the general wave theory was itself eventually given up), but they are *relatively* speaking permanent, and this higher degree of relative entrenchment has played a big role in the success of science.

4. The 'Darwinian' Analogy: what was Popper's *real* view about 'conjectures'?

Is there any inkling of a more sophisticated view about rationally analysable theory-construction in Popper's work?

Or is the impression justified that he flatly denies the possibility of any such rational analysis? Certainly it is easy to point to passage after passage that seems unambiguously to support the 'flat denial' interpretation.

There is, of course, the famous remark in *Logik der Forschung* that

> The initial stage, the act of conceiving or inventing a theory, seems to me neither to call for logical analysis nor to be susceptible of it. The question how a new idea occurs to a man—whether it is a musical theme, a dramatic conflict, or a scientific theory—may be of great interest to empirical psychology; but it is irrelevant to the logical analysis of scientific knowledge. (*LSD*, p. 31)

But even earlier, in his *Die Beiden Grundprobleme der Erkenntnistheorie*, Popper had espoused an explicitly 'Darwinian' account of theory-production, a view which seems directly at odds with what might be called the rational heuristic view outlined in section 3. In that early work, Popper asserted:

> there exists no law-like dependence between receptions, between new objective conditions and the emergence of reactions (or rather there is only one form of dependence, namely the selective one, which renders non-adaptive reactions worthless...)[11]

---

[11] Popper, *BG*. This was written before *Logik der Forschung* but finally published only in 1979. The quotation is from p. 27 and the translation is due to Elie Zahar.

(Popper here clearly means by 'receptions' the evaluation of a theory once it has been articulated, and by 'reactions' the articulation of a new theory.) Popper repeated this claim at various points throughout his career. As we have seen, he explicitly asserted, for example (*OK*, p. 144), that

> The growth of knowledge—or the learning process—is not ... a cumulative process but one of error-elimination. It is Darwinian selection, rather than Lamarckian instruction.

In his 1973 Herbert Spencer Lecture, revealingly entitled 'The Rationality of Scientific Revolutions. Selection *versus* Instruction', he again argued what appears at least to be an unequivocal selectionist, anti-instructionist line. Popper was there concerned to draw parallels between biological adaptation, behavioural adaptation and the 'adaptation' of scientific theories to their epistemic environment. He asserted that:

> On all three levels—genetic adaptation, adaptive behaviour, and scientific discovery—the mechanism of adaptation is fundamentally the same... If mutations or variations or errors occur, then there are new instructions, which also arise *from within the structure*, rather than *from without*, from the environment.
>
> These inherited structures are exposed to certain pressures, or challenges, or problems: to selection pressures; to environmental challenges; to theoretical problems. In response, variations of the genetically or traditionally inherited *instructions* are produced, by methods that are at least partly *random*. On the genetic level, these are mutations and recombinations of the

coded instructions...; on the scientific level, they are new
and revolutionary tentative theories... It is important
that these tentative trials are changes that originate
within the individual structure in a more of less random
fashion—on all three levels. The view that they are not
due to instruction from without, from the environment,
is supported (if only weakly) by the fact that very similar
organisms may sometime respond in very different ways
to the same new environmental challenge. (*MF*,
pp. 78–79)

It is true that Popper here stressed that genetic mutations
occur against the background of an otherwise non-mutating
genome which is systematically transmitted by inheritance.
And it might be claimed on his behalf that this provides the
analogue for the material that gets transmitted from older to
newer theory in a systematic way. It is unclear that this
analogy can be made to work in any even remotely precise
sense, but it is surely abundantly clear that it is not generally
by 'random variation', *even against a fixed background*, that
scientists produce a multitude of theories which are then let
out into the critical jungle to see if they survive.[12] The way
for example that Fresnel produced the transverse wave
theory of light is, as we saw, *clearly* 'instructed by the
[epistemic] environment'. Fresnel made a deliberate and
conscious attempt to produce a theory that would fill an
existing epistemic environmental niche—that is, a theory

---

[12] I should not want to assert that scientists are never reduced to 'random
conjecture' but this is both unusual and very much scraping the
heuristic barrel.

that would preserve the successes of its predecessor while solving the empirical problems faced by its predecessor. Nothing could be less Darwinian.

Further evidence that Popper was committed to denying any role for rationally analysable heuristics can be gleaned from his reaction to Kuhn's idea of 'normal science'. Popper interpreted Kuhn's emphasis on the importance of normal science as amounting to the advocacy of some form of rationally unmotivated dogmatism. Popper admitted that 'Kuhn has discovered something which I failed to see ... what he has called "normal science" and the "normal scientist"' (*MF*, p. 57). But what Kuhn discovered, though real, Popper found deeply disturbing: 'In my view, the "normal" scientist, as Kuhn describes him, is person one ought to be sorry for' ('NSD', p. 52). And Popper suggested that, were all scientists to become 'normal scientists', this would be 'the end of science as we know it' ('NSD', p. 57).

Popper's remark stems from his horror of dogmatism. But, while there no doubt are passages where Kuhn overdoes the need for scientists to be 'committed' to their paradigm-constituting general theories, and while (as Kuhn in effect himself later admitted) there is no doubt that he greatly overdoes the 'paradigm monopoly' view, the point underlying his analysis is of course *not* that scientists should be 'dogmatic' in any obviously unacceptable sense. Instead he is arguing that the history of science shows that scientific progress is best made (perhaps only made) *not* by holding every assumption equally open to criticism in the sort of critical free-for-all that Popper *seemed* to advocate, but by a process in which background theories are taken as relatively well-entrenched, and are systematically used along with previous successes in science *in the construction of new specific theories.* (Of course, even these relatively well-entrenched background principles may *eventually* themselves be rejected—in so-called revolutions.)

A scientist who holds on to Newton's theory in the light of the Uranian 'anomalies' is being no more 'dogmatic' than his colleague who regards those anomalies as pointing to the falsity of Newton's theory. The two are simply placing the 'blame' for the refutation of the same *overall* system in different places: each accepts certain theories (in the latter case, the auxiliary and instrumental theories rather than the core) and therefore finds the evidence a reason to reject others. Kuhn, when properly construed, is simply recording the fact about the history of science that, once a theory such as Newton's has proved successful, it has generally proved fruitful to regard it as relatively well-entrenched and to look to deal with anomalies for the overall system by taking that 'central' theory—*relatively*—for granted. This, in turn, is no unmotivated choice, instead it allows the scientist to take advantage of various heuristic ideas based on the central theory.[13]

It was in connection with Kuhn's vision of long periods of essentially cumulative 'normal' development, punctuated by occasional 'revolutions' that Popper

[13] This, incidentally, is why Popper's concession that 'dogmatism' may occasionally have *some* value (see 'NSD', p. 55 and *MF*, p. 16) is off-beam. There is *never* any need for 'dogmatism', only a need for good ideas about which particular parts of large theoretical systems need to be amended in view of experimental difficulties.

155

countered that 'science is revolution in permanence'. I believe that, ironically, the correct criticism of Kuhn's account is not that he underestimated the revolutionary nature of normal science but that he underestimated the normality of so-called revolutions. It would be easy to get involved here in essentially semantic squabbles about what counts as a revolutionary change—especially if, like Popper (and also—though for different reasons—like Kuhn in the *Postscript*), we allow ourselves to talk in effect about 'mini-revolutions'. I have already given one example, however, of a seemingly quite radical change in theory—Fresnel's switch from the fluid to the solid theory of the ether—that certainly felt like a big change to the protagonist and his contemporaries. It seemed strongly counterintuitive that the ether could be a solid and yet let the planets move through it with no perceptible effect. Yet this shift, as we saw, involved nothing like a 'conversion experience' and nothing like a 'bold conjecture', but instead a systematic argument on the basis of what was then considered known *plus* Fresnel and Arago's new phenomena.

A defender of Kuhn would no doubt argue that this particular change, though perhaps more radical than it might initially seem, is nonetheless not radical *enough* to count as revolutionary in his terms—that the shift from longitudinal to transverse waves is part of 'normal science'. But it is easy enough to point to cases of theory-shifts that Kuhn explicitly counts as revolutionary, and that fit exactly this same model—the only difference being that the background 'premises' involved were of a still more general kind. Kuhn explicitly includes, for example, the shift from the

Newtonian particulate theory of light (which, roughly speaking, was the most widely accepted fundamental theory of light in the eighteenth century) to Fresnel's wave theory of light (which had certainly become the most widely accepted such theory by the mid-1830s) as a scientific revolution. Yet the basic wave theory too was not arrived at by anything remotely describable as a 'religious conversion' or as a 'blind conjecture' following the refutation of the earlier theory. Rather the wave theory itself could be argued for by something like a 'deduction from the phenomena'—an argument which Fresnel himself hinted at more than once. (The argument is found explicitly in Huygens's *Treatise on Light*.)

A more fundamental part of background knowledge in the early nineteenth century than any claim about light was what might be called the 'classical world view'—the theory that all physical processes basically involve matter in motion under the action of forces. What, given this world view, might light sources emit? The general 'classical' theory (already incorporated into a range of successful specific theories) permitted only two possibilities: light sources like the sun either emit matter or they, so to speak, emit motion (or, of course, a combination). Basic ideas about matter split the first possibility into two—light might consist of a continuous stream of matter or of a discontinuous stream of particles. The former ran into a range of empirical difficulties and so, by Fresnel's time, had the particulate theory: theoretical systems based on it had done no more than accommodate various known phenomena and even that accommodation had been achieved only at the expense of a range of assumptions of extreme implausibility. But if matter was out, background knowledge also implied that disembodied motion was a nonsense: the motion had to be held by some medium in the finite interval (again the finite velocity of light was a 'given' established part of background knowledge) between leaving a source and meeting a receptor. Hence there must exist some material medium between the source and the receptor which carries the disturbances constituting light. Since light is known to be transmitted freely through a vacuum, that medium cannot be the air. Finally it was again part of accepted background knowledge that light, whatever it might precisely be, must be fundamentally periodic: a monochromatic light beam must, somehow or other, re-exhibit the same property at regular intervals. Thus we finally have the theory of light as periodic disturbances in an all-pervading, intangible, material medium—that is, we have the fundamental general idea of the classical wave theory.

What otherwise might appear as a bold conjecture—that there exists an invisible, intangible medium filling the whole of space, vibrations in which constituted light—is thus shown to be the entirely rationally reconstructible result of plugging new data and judgments based on data about other possible theories into background knowledge.

Elie Zahar has argued that, once we take into account the heuristic use of the correspondence principle, then even what seem to be *very* 'revolutionary' revolutions (such as the relativistic one) can be given gradualistic explanations along the above lines.[14] The 'correspondence

---

[14] See his 'Logic of Discovery or Psychology of Invention?' *British Journal for the Philosophy of Science* (1984), pp. 243–261.

principle' in this sense is the requirement that, in cases of theory change, the new theory, if it is to gain scientific acceptance, must always share the empirical successes of the old theory—a feat that the new theory generally achieves by 'going over' (*via* some mathematically characterized limiting process) to the old in the empirical domain in which the old was successful. As a requirement for the acceptance of a theory 'already on the table'; this principle has often been stressed—notably by Popper himself in several places. What Zahar has shown, I believe, is that the principle is often quite deliberately and consciously used in the *construction* of the new theory. That is, it is used as a heuristic principle rather than merely as an *ex post* appraisal criterion.

So far, then, so unambiguous: the articulation of promising new theories is *not* (unsurprisingly) a question of throwing out possible conjectures 'at random' and then subjecting them to rigorous selection pressure; the process is not (even approximately) analogous to Darwinian natural selection; and yet Popper time and again emphasized this alleged analogy. He seems, therefore, consistently to have held the wrong view about theory-production and to have misinterpreted accounts that might have pointed him in the right direction. As in the case of refutations, however, it is not too difficult to find remarks that might be used to support the claim that he held a more sophisticated view.

For example, he more than once insisted that in order for a scientist to produce a worthwhile conjecture, he must be 'fully immersed in the problem-background.' It would, of course, be quite mysterious why this should be necessary if 'the only form of dependence between [the production of a new theory and the epistemic situation] were the selective one...' But Popper seems never to have followed up this hint and seems to be have remained unaware of the mystery.

Secondly, Popper often stressed that metaphysics could exert a beneficial influence on the development of science. This position is one that, he repeatedly insisted, most clearly marked him out from the logical positivists. He claimed to have invented the idea of a 'metaphysical research programme' (an idea subsequently appropriated, but misinterpreted by Lakatos). Such metaphysical programmes may 'play a crucial role in the development of science'. General background claims such as those of mechanism or determinism that, I have argued, have figured as 'premises' in the deduction of theories 'from the phenomena' are of course reasonably regarded as metaphysical. However, Popper, whose attitude to such 'programmes' is by no means unambiguously positive, never seems to have developed this idea and certainly never spelled out how *exactly* metaphysical principles play a 'crucial role' in the development of fully scientific theories. Although he complained that Lakatos took the idea from him, there is no real hint that Popper was really aware of the central idea in Lakatos's conception: that of an articulable 'positive heuristic' providing definite guidance for the construction of theories within a programme.

The *third* sort of reason why Popper's view of theory-production is less clear (and therefore less clearly wrong) than might at first be supposed is that his later developments of the 'Darwinian' account of theory-production are cast around with qualifications and asides that often do not seem to cohere with the central message. It seems difficult indeed to hold the view that theorizing has no element of goal-directedness or to hold that the superabundance of rival theories ('mutations') really exists that would be necessary to give any plausibility to a properly selectionist account. (Scientists of course generally find it extremely difficult to produce *one* theory that solves the problems a fully acceptable theory in a given area would need to solve, let alone a superabundance of them.) Not surprisingly, then, there are hints that Popper saw some of the difficulties here. But rather than simply give up the Darwinian analogy on account of those difficulties, he seems to have reacted by asserting it all the more forcefully as the *fundamentally* correct view, while adding what appear to be details that, taken together, and in so far as they are clear at all, seem almost to amount to surrendering the view. The basic message, as we saw, is that Popper is on the side of 'selection' rather than 'instruction' both in biology and in the case of theory-production in science. But then certain detailed remarks suggest that he interprets 'selection' in a way that is, to say the least, rather unorthodox. So, for example, Popper introduced an alleged distinction between 'blind' and 'random' variations, and allowed that while variations in scientific discovery are 'more or less' (sic) random, they are 'not completely blind' (*MF*, p. 81). And he finally seemed to blur the whole distinction between Darwinian and Lamarckian approaches by claiming that *even in the properly biological case* 'if there were no variations, there could not be adaptation or evolution; and so we can say that the occurrence of mutations is either partly controlled by the need for them, or functions as if it was' (*MF*, p. 79).

I have no doubt that, given sufficient motivation, a case could be constructed on the basis of such remarks that Popper had a more sophisticated view of theory-production than the bold conjectures view his more general remarks seem to recommend. But, as Popper himself has argued in other connections, it may be better to hold a view that is clearly wrong, rather than one that escapes being clearly wrong only by virtue of not being clear.

Again, I thankfully leave further exegetical research to others and settle for a qualified judgment: *if* there are hints of a more sophisticated account of theory-production in Popper's work then, unlike in Kuhn and Lakatos, there is no attempt to develop them.

Popper is famous for his strikingly simple view of science. Unfortunately, despite its undoubted charms, the view is much too simple to be true. No doubt at a sufficiently high level of generality, science can be described as a process of trial and error, conjecture and refutation. But if it is to be so described, then both the process of 'conjecturing' and that of identifying 'error' need to be understood in quite sophisticated ways. Sometimes Popper seems to be aware of this and sometimes he seems vigorously to deny it; but in any event others have certainly managed to describe the processes with much greater clarity, accuracy and detail.

5 Popper's Contribution to the Philosophy of Probability

DONALD GILLIES

1. Introduction

Popper's writings cover a remarkably wide range of subjects. The spectrum runs from Plato's theory of politics to the foundations of quantum mechanics. Yet even amidst this variety the philosophy of probability occupies a prominent place. David Miller once pointed out to me that more than half of Popper's *The Logic of Scientific Discovery* is taken up with discussions of probability. I checked this claim using the 1972 6th revised impression of *The Logic of Scientific Discovery*, and found that of the approximately 450 pages of text, approximately 250 are to do with probability. Thus Miller's claim is amply justified. It seems indeed that the philosophy of probability was one of Popper's favourite subjects, and, as we shall see, Popper certainly enriched the field with several striking innovations. In this area, as in others, Popper held very definite views, and criticized his opponents in no uncertain terms. Popper was an objectivist and anti-Bayesian, and his criticisms were directed against subjectivism and Bayesianism.

As well as carrying out his own research in the philosophy of probability, Popper stimulated interest in the subject among members of his department at the London School of Economics. When I arrived there as a graduate student in 1966, Imre Lakatos was editing a volume on one part of the subject (*The Problem of Inductive Logic*), and himself wrote a paper for the volume.[1] Imre Lakatos did not return to the subject, but this paper is itself a notable contribution. Among the graduate students who were my contemporaries, several became interested in the philosophy of probability, and went on to do research in the subject. Apart from me, these included: Peter Clark, Colin Howson, David Miller and Peter Urbach. David Miller and I have, on the whole, been sympathetic to Popper's approach to the subject. Peter Clark might be classified as neutral. He has worked on the nature of probability in statistical mechanics, and, in a paper from 1987 argues against Popper that there can be physical probabilities in a deterministic world.[2] On the other hand he agrees with Popper that such probabilities are objective rather than subjective. Colin Howson and Peter Urbach are more strongly opposed to Popper, and have adopted the very position which Popper attacked (subjective Bayesianism). Their book *Scientific Reasoning: The Bayesian Approach*[3] contains both a development of this position, and many sharp criticisms of Popper's ideas on the subject. It is in my view quite a tribute to Popper and his department at LSE that it was able to generate critics as well as followers. In

[1] I. Lakatos, 'Changes in the Problem of Inductive Logic', in I. Lakatos (ed.), *The Problem of Inductive Logic* (North Holland, 1968), pp. 315–417.

[2] P. Clark, 'Determinism and Probability in Physics', *The Aristotelian Society, Supplementary Volume*, **LXI** (1987), pp. 185–210.

[3] Open Court, 1989.

accordance with the Popperian principle that criticism is good, I will, in what follows, not only expound Popper's ideas, but also give criticisms of these—particularly the criticisms proposed by Howson and Urbach. Sometimes Popper seems to me to have the better of the argument, but at other times victory has to be conceded to Howson and Urbach. There is even a third situation which occurs. Howson and Urbach pose a serious problem for Popper. This can, in my view, be solved, but only by using methods, of which Popper himself might not have approved.

Since Popper wrote so extensively on the philosophy of probability, I cannot discuss all aspects of his work in this area. What follows is frankly a personal selection of what seem to me to be his most important ideas on the subject. I have divided these into three topics which are expounded in the next three sections of the paper. Section 2 deals with the propensity theory; section 3 with the application of falsifiability to probability; and section 4 with the claim that corroboration is not a probability function.

2. The Propensity Theory

In his 1934 *The Logic of Scientific Discovery*, chapter VIII, Popper discusses the interpretation of probability in scientific theories. Characteristically he argues for an objective interpretation, and, more specifically develops a version of the frequency theory which had been advocated by van Mises.[4]

---

[4] See R. von Mises, *Probability, Statistics and Truth* (1928; second revised English edition, Allen and Unwin, 1951).

Somewhat unusually Popper was here in agreement with the Vienna Circle, most of whom supported the frequency theory. Later on, however, Popper came to the conclusion that he had been mistaken on this point, and accordingly developed a new interpretation of probability, which was still objective, but differed from the frequency theory. This was his *propensity theory of probability*, which constituted a definite innovation in the field. Popper published an account of the new theory in his 1959 article, ('PIP'), and gave a fuller exposition of it in Part II (pp. 281–401) of *RAS*. As a matter of fact this book had already been written by 1959, but was only published twenty-four years later, in 1983.

The problem which gave rise to the propensity theory had already been considered by Popper in 1934. The question was whether it was possible to introduce probabilities for single events, or *singular probabilities* as Popper calls them. Von Mises, assuming of course his frequency theory of probability, denies that such probabilities can validly be introduced. The example he considers is the probability of death. We can certainly introduce the probability of death before 80 in a sequence of say 40-year-old English non-smokers. It is simply the limiting frequency of those in the sequence who die before 80. But we can consider the probability of death before 80 for an individual person (Mr Smith say)? Von Mises answer is: 'no!':

> We can say nothing about the probability of death of an individual even if we know his condition of life and health in detail. The phrase 'probability of death', when it refers to a single person has no meaning at all for us. This

> is one of the most important consequences of our
> definition of probability...[5]

Of course it is easy to introduce singular probabilities on the subjective theory, according to which probabilities are the degrees of belief of particular individuals in some event, and are measured by the rate at which these individuals would bet on the event under specified circumstances. All Mr Smith's friends could, for example, take bets on his dying before 80, and hence introduce subjective probabilities for this event. Clearly, however, this procedure would not satisfy an objectivist like Popper. The key question for him was whether it was possible to introduce objective probabilities for single events.

Von Mises denied that there could be objective singular probabilities, but Popper disagreed, partly because he wanted such objective singular probabilities for his interpretation of quantum mechanics. Von Mises had suggested that the probability of an event was its limiting frequency in a long sequence of events, which he called a *collective*. Popper considered a single event which was a member of one of von Mises' collectives, and made the simple suggestion that its singular probability might be taken as equal to its probability in the collective as a whole. In 'PIP' Popper considers an objection to this earlier view of his, and this leads him to his new theory of probability.

Popper's argument is as follows. Begin by considering two dice: one regular, and the other biassed so that the

[5] Ibid., p. 11.

probability of getting a particular face (say the 5) is 1/4. Now consider a sequence consisting almost entirely of throws of the biassed die but with one or two throws of the regular die interspersed. Let us take one of these interspersed throws and ask what is the probability of getting a 5 on that throw. According to Popper's earlier suggestion this probability must be 1/4 because the throw is part of a collective for which prob(5) = 1/4. But this is an intuitive paradox, since it is surely much more reasonable to say that prob(5) = 1/6 for any throw of the regular die.

One way out of this difficulty is to modify the concept of collective so that the sequence of throws of the biassed die with some throws of the regular die interspersed is not a genuine collective. The problem then disappears. This is just what Popper does.

> All this means that the frequency theorist is forced to introduce a modification of his theory—apparently a very slight one. He will now say that an admissible sequence of events (a reference sequence, a 'collective') must always be a sequence of repeated experiments. Or more generally he will say that admissible sequences must be either virtual or actual sequences which are characterized by a set of generating conditions—by a set of conditions whose repeated realization produces the elements of the sequences. ('PIP', p. 34)

He then continues a few lines later:

> Yet, if we look more closely at this apparently slight modification, then we find that it amounts to a transition from the frequency interpretation to the propensity interpretation.

168

In this interpretation the generating conditions are considered as having a propensity to produce the observed frequencies. As Popper puts it:

> But this means that we have to visualize the conditions as endowed with a tendency or disposition, or propensity, to produce sequences whose frequencies are equal to the probabilities; which is precisely what the propensity interpretation asserts. ('PIP', p. 35)

But does Popper really succeed in introducing objective singular probabilities in a satisfactory fashion? There is a problem with such probabilities which was discussed by Ayer,[6] and which could be called the 'conditionality difficulty'. Suppose we are trying to assign a probability to a particular event, then the probability will vary according to the set of conditions which the event is considered as instantiating—according, in effect, to how we describe the event. But then we are forced to consider probabilities as attached primarily to the conditions which describe the event and only secondarily to the event itself.

Consider, for example, the probability of a certain man aged 40 living to be 41. Intuitively the probability will vary depending on whether we regard the individual merely as a man or more particularly as an Englishman; for the life expectancy of Englishmen is higher than that of mankind as a whole. Similarly the probability will alter depending on whether we regard the individual as an Englishman aged

---

[6] See A. J. Ayer, 'Two Notes on Probability', in *The Concept of a Person and Other Essays* (Macmillan, 1963), pp. 188–208.

40 or as an Englishman aged 40 who smokes two packets of cigarettes a day, and so on. This does seem to show that probabilities should be considered in the first instance as dependent on the properties used to describe an event rather than as dependent on the event itself.

Howson and Urbach take up this criticism and use it to argue that single case probabilities are subjective rather than objective. However they also suggest that singular probabilities, though subjective, may be based on objective probabilities. Suppose, for example, that the only relevant information which Mr B has about Mr A is that Mr A is a 40-year-old Englishman. Suppose Mr B has a good estimate (p say) of the objective probability of 40-year-old Englishmen living to be 41. Then it would be reasonable for Mr B to put his subjective betting quotient on Mr A's living to be 41 equal to p, and thereby making his subjective probability objectively based. This does not, however, turn Mr B's subjective probability into an objective one, for, consider Mr C, who knows that Mr A smokes two packets of cigarettes a day, and who also has a good estimate of the objective probability (q say) of 40-year-old Englishmen who smoke two packets of cigarettes a day living to be 41. Mr C will put his subjective probability on the same event (Mr A living to be 41) at a value q different from Mr B's value p. Once again the probability depends on how the event is categorized rather than on the event itself. Howson and Urbach put the point as follows:

> single-case probabilities ... are not themselves objective. They are subjective probabilities, which considerations of consistency nevertheless dictate must be set equal to the

objective probabilities just when all you know about the single case is that it is an instance of the relevant collective. Now this is in fact all that anybody ever wanted from a theory of single-case probabilities: they were to be equal to objective probabilities in just these conditions. The incoherent doctrine of objective single-case probabilities arose simply because people failed to mark the subtle distinction between the values of a probability being objectively based and the probability itself being an objective probability.[7]

This criticism of Howson and Urbach seems to me perfectly fair, and I do not think objective singular probabilities can be introduced in a satisfactory fashion. Does this mean that I therefore reject the propensity theory of probability? Not at all. Popper introduced the theory because of the problem of singular probabilities, but it turns out that there are other quite different reasons why such a theory is desirable.

It is a consequence of von Mises's frequency theory that probabilities ought only to be introduced in physical situations in which we have an empirical collective, that is a long sequence of events or entities. If, however, we adopt Popper's propensity theory, it is quite legitimate to introduce probabilities relative to a set of conditions *even though these conditions are not repeated a large number of times*. To say: 'The probability of outcome A of conditions S = 1/6' is equivalent, on Popper's approach, to saying: 'The conditions S have a tendency or disposition or propensity such that, if

[7] C. Howson and P. Urbach, *Scientific Reasoning. The Bayesian Approach* (Open Court, 1989), p. 228.

they were to be repeated infinitely often, a random sequence of attributes would result in which attribute A would have the limiting frequency 1/6'. As the random sequence here is hypothetical, we do not have to consider it as corresponding to a long sequence given in experience. We would be allowed to postulate probabilities (and might even obtain testable consequences of such a postulation) when the relevant conditions were only repeated once or twice. Thus the propensity theory broadens the range of situations in which it is legitimate to introduce objective probabilities. This is a significant step forward, even if the broader range does not include singular probabilities.

This point is connected to another one. Von Mises's frequency theory of probability was strongly and consciously based on Mach's operationalist and positivist philosophy of science.[8] Thus von Mises's definition of probability as limiting frequency is intended as an operational definition of a theoretical concept (probability) in terms of an observational one (frequency). But Popper in *The Logic of Scientific Discovery* is highly critical of Machian positivism. It turns out, therefore, that he was wrong in that work to agree with the majority of the Vienna Circle in adopting the frequency theory, since the positivism underpinning that theory was very different from his own views about science. By introducing a new objective, but non-frequency, theory of probability, Popper was creating an interpretation of probability

---

[8] A justification of this statement with extensive quotations from both Mach and von Mises is to be found in my *An Objective Theory of Probability* (Methuen, 1973), pp. 1−7 and 37−47.

which agreed much better with his own philosophy of science. In his new propensity approach, the theoretical concept of probability is not defined in terms of observables, but is introduced as an undefined term which is connected indirectly with observation. This point of view obviously accords well with Popper's general views on philosophy of science, but, at the same time, it raises the question of how exactly the theoretical term (probability) is linked to the observable term (frequency). It turns out that the solution to this problem lies in the consideration of another problem which Popper posed for the philosophy of probability. This is the question of how falsifiability applies to probability, and I will discuss it in the next section.

3. The Application of Falsifiability to Probability

Having advocated falsifiability in *The Logic of Scientific Discovery*, and having also a considerable interest in probability, it was very natural for Popper to consider how falsifiability applies to probability, and this is just what he does do in chapter VII of his famous work. It turns out that there is a difficulty connected with the falsifiability of probability statements which Popper himself states very clearly as follows:

> The relations between probability and experience are also still in need of clarification. In investigating this problem we shall discover what will at first seem an almost insuperable objection to my methodological views. For although probability statements play such a vitally

important role in empirical science, they turn out to be in principle *impervious to strict falsification*. Yet this very stumbling block will become a touchstone upon which to test my theory, in order to find out what it is worth (*LSD*, p. 146).

To see why probability statements cannot be falsified, let us take the simplest example. Suppose we are tossing a bent coin, and postulate that the tosses are independent and that the probability of heads is p. Let prob(m/n) be the probability of getting m heads in n tosses. Then we have

$$\text{prob}(m/n) = {}^{n}C_{m}p^{m}(1 - p)^{n-m}$$

So, however long we toss the coin (that is, however big n is) and whatever number of heads we observe (that is, whatever the value of m), our result will always have a finite, non-zero probability. It will not be strictly ruled out by our assumptions. In other words, these assumptions are 'in principle *impervious to strict falsification*'. Popper's answer to this difficulty consists in an appeal to the notion of methodological falsifiability. Although, strictly speaking, probability statements are not falsifiable, they can none the less be used as falsifiable statements, and in fact they are so used by scientists. He puts the matter thus: 'a physicist is usually quite well able to decide whether he may for the time being accept some particular probability hypothesis as "empirically confirmed", or whether he ought to reject it as "practically falsified"...' (*LSD*, p. 191).

Popper's approach has been strongly vindicated by standard statistical practice. Working statisticians are constantly applying one or other of a battery of statistical tests.

Now, whenever they do so, they are implicitly using probability hypotheses, which from a strictly logical point of view are unfalsifiable, as falsifiable statements. The procedure in any statistical test is to specify what is called a 'rejection region', and then regard the hypothesis under test (H say) as refuted if the observed value of the test statistic lies in this rejection region. Now there is always a finite probability (called the 'significance level' and usually set at around 5%) of the observed value of the test statistic lying in the rejection region when H is true. Thus H is regarded as refuted, when, according to strict logic, it has not been refuted. This is as much as to say that H is used as a falsifiable statement, even though it is not, strictly speaking, falsifiable, or, to put the same point in different words, that methodological falsifiability is being adopted.

The first important statistical tests were introduced in the period 1900–1935 by Karl Pearson (the chi-square test), Student (W. S. Gosset) (the t-test), and R. A. Fisher (who introduced the F-test, as well as improving the two preceding tests). These tests are still widely used today, though others have of course been devised subsequently. Now the interesting thing is that statistical tests were introduced and came to be very widely adopted quite independently of Popper's advocacy of methodological falsifiability. Statistical tests are, however, based implicitly on methodological falsifiability, and their introduction and widespread adoption by statisticians provides striking corroboration of the value of Popper's approach.

But what do Popper's critics Howson and Urbach say about this success? They do quite frankly acknowledge its existence, but they claim that this success is undeserved, and that statisticians should adopt different methods. Thus they write:

> it is fair to say that their theories, especially those connected with significance testing and estimation, which comprise the bulk of so-called classical methods of statistical inference, have achieved pre-eminence in the field. The procedures they recommended for the design of experiments and the analysis of data have become the standards of correctness with many scientists.
>
> In the ensuing chapters we shall show that these classical methods are really quite unsuccessful, despite their influence amongst philosophers and scientists, and that their pre-eminence is undeserved.[9]

Howson and Urbach go on in chapter 5 of their book to criticize the standard approach to statistical testing in detail, and even to argue that the tests normally used are unsatisfactory. Thus they present some criticisms of the chi-square test, and conclude 'What is needed instead is a test based on reasonable principles, and the present problem suggests this is beyond the scope of the chi-square test. It should be discarded.'[10]

I have already discussed and tried to answer Howson and Urbach's detailed criticisms of the standard approach to statistical testing,[11] and I will not repeat these arguments here. Suffice it to say that I think it very unlikely

[9] Howson and Urbach, *Scientific Reasoning*, p. 11.   [10] Ibid., p. 136.
[11] See my 'Bayesianism versus Falsificationism', *Ratio*, 3 (1990), pp. 82–98, at pp. 90–98.

indeed that a test which is so widely and successfully used as the chi-square test will be discarded by the statistical community. So on this point it seems to me that Popper has the better of the argument.

It remains to show that, if we adopt methodological falsifiability, we can solve the problem of linking objective undefined probabilities (propensities) to observed frequencies. The idea of methodological falsifiability is that, although probability statements are not strictly speaking falsifiable, they should be used in practice as falsifiable statements. If we adopt this position, we ought to try to formulate what could be called a *falsifying rule for probability statements* which shows how probability statements should be used as falsifiable statements. Such a rule should obviously agree with and implicitly underlie the practice of statistical testing. Elsewhere, I have tried to formulate such a rule.[12] I will here try to state the main idea without going into full technical detail.

Let H be a statistical hypothesis. Suppose from H it can be deduced that a random variable X has a bell-shaped distribution D of roughly the form shown in Figure 1. The two points a and b are chosen so that D is divided into a 'head', i.e. $b \geq X \geq a$, and tails, i.e. $X < a$ or $X > b$. The tails are such that the probability of obtaining a result in the tails, given H, has a low value known as the significance level. The significance level is normally chosen between 1% and 10%,

---

[12] See my 'A Falsifying Rule for Probability Statements', *British Journal for the Philosophy of Science*, **22** (1971), pp. 1–22, and *An Objective Theory of Probability* (Methuen, 1973), pp. 161–226.

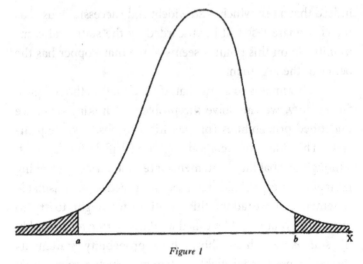

Figure 1

Figure 1

5% being the most common value. Suppose further that X is a test statistic, that is to say that X is a function of the observed data whose value can be calculated from the data. Our falsifying rule for probability statements now states that if the value obtained for X is in the tails of the distribution, this should be regarded as falsifying H; while, if the value of X is in the head of the distribution, this should be regarded as corroborating H.

Broadly speaking this falsifying rule agrees with the practical procedures adopted when such standard statistical tests as the chi-square, the t-test, or the F-test are applied. There are indeed some problems connected with 1-tailed as opposed to 2-tailed tests, and with the Neyman paradox. These are rather technical matters, however, and the interested reader is referred to the fuller and more mathematical treatments of the question referred to in footnote 12. The point I want to make here is that the adoption of a falsifying rule such as the one just given establishes a link between undefined propensities and observed frequencies, and this can be illustrated by simple coin tossing examples.

Suppose we toss a coin n times, and observe m heads. If we postulate that the tosses are independent, and that the probability of heads on each toss is p, then it follows (by a theorem first proved by De Moivre in 1733) that the distribution of m/n is approximately normal with mean p and standard deviation $\sqrt{[p(1-p)/n]}$. If now, in accordance with our falsifying rule, we cut off the tails of this normal distribution at the 5% level of significance, our hypothesis is confirmed if we have

$$p + 1.96\sqrt{[p(1-p)/n]} \geq m/n \geq p - 1.96\sqrt{[p(1-p)/n]}$$

and falsified if m/n lies outside that interval.

Let us now consider the special case when $p = 0.5$. Applying our falsifying rule at the 5% significance level, we, in effect, predict that it is practically certain that m/n will lie in the interval $\pm 0.98/\sqrt{n}$, and regard our underlying hypothesis as refuted if the observed value of m/n lies outside this interval.

To see how this applies in practice, I give in Table 1 the results of some coin tossing experiments. The first was performed by me, the second by Buffon, and the third and fourth by Karl Pearson. In each case I give the allowable deviation around 0.5 as calculated by the falsifying rule at the 5% level of significance, and also the observed relative frequency of heads. As can be seen, the results of all

Table 1. *Some Coin Tossing Experiments*

Author	Number of Tosses	Allowable Deviation	Observed Relative Frequency of Heads
Gillies	2,000	±0.022	0.487
Buffon	4,040	±0.015	0.507
K. Pearson	12,000	±0.009	0.502
K. Pearson	24,000	±0.006	0.501

4 experiments confirmed the hypothesis of an unbiassed coin and independent tosses. These results show in a vivid way that, even if probability is not defined in terms of frequency, the adoption of a falsifying rule for probability statements can establish a link between probability and observed frequency.

So far I have discussed Popper's analysis of probability as it appears within theories of the natural sciences, for example in physics or biology. However an attempt is sometimes made to use probability to assess scientific theories relative to evidence. To try to estimate, for example, the probability of a theory or law relative to some given evidence. Popper has also had much of interest to say about this branch of the philosophy of probability, and I will examine some of his views on this subject in the next section.

4.   The Claim the Corroboration is not a Probability Function

In science and everyday life, we use the notion of evidence (*e*) *corroborating* a hypothesis (*h*) or a prediction (*d*). The theory of corroboration is an attempt to analyse the crucial notion of corroboration. Now it is common to use the phrases: '*e* corroborates *h*' and '*e* renders *h* probable' as synonyms, and it is therefore natural to identify corroboration with probability. Indeed many authors have implicitly assumed the identity of the two concepts without discussing the matter at all. It is one of Popper's most striking innovations in the philosophy of probability that he challenged this implicit assumption and argued strongly against it. Even those who disagree with Popper on this point have been affected by his work, for they now have to argue for the assumption that corroboration is probability, an assumption which was before made without argument.

Popper's point is that the calculus of probabilities was developed in connection with the study of chance. It was thus designed for statistical probabilities and it is by no means obvious that degrees of corroboration will obey the same formal rules. Nonetheless writers previous to Popper often assumed the identity of corroboration and probability more or less automatically, and without considering alternative possibilities. A nice example of this is provided by the following passage from Poincaré's *Science and Hypothesis*:

> in the two preceding chapters I have several times used the words 'probability' and 'chance.' 'Predicted facts,' as I said above, 'can only be probable'. However solidly founded a prediction may appear to be, we are never absolutely certain that experiment will not prove it false; but the probability is often so great that practically it may be accepted. And a little farther on I added:- 'See

what a part the belief in simplicity plays in our
generalizations. We have verified a simple law in a large
number of particular cases, and we refuse to admit that
this so-often-repeated coincidence is a mere effect of
chance.' Thus, in a multitude of circumstances the
physicist is often in the same position as the gambler
who reckons up his chances. Every time that he reasons
by induction, he more or less consciously requires the
calculus of probabilities and that is why I am obliged ...
to interrupt our discussion of method in the physical
sciences in order to examine a little closer what this
calculus is worth, and what dependence we may place
upon it.[13]

Poincaré starts here by considering predictions which are
well supported by evidence. He goes on to well-founded
generalizations. Thence he passes to 'the gambler who
reckons up his chances' and so to the calculus of probabilities. But is the calculus developed for the gambler necessarily the appropriate tool for the physicist considering the
degrees of corroboration of various hypotheses? Poincare
assumed almost unconsciously that it was, and he was
followed in this by Keynes, Jeffreys, and Carnap.[14] This
assumption was, however, challenged by Popper.

[13] H. Poincaré, *Science and Hypothesis* (1902, English translation, Dover,
1952), pp. 183–184.
[14] See J. M. Keynes, *A Treatise on Probability* (Macmillan, 1921); H.
Jeffreys, *Theory of Probability* (Oxford University Press, 1939); R.
Carnap, *Logical Foundations of Probability* (University of Chicago
Press, 1950).

Let us write C(*h,e*) for the degree of corroboration of *h* given *e*, and P(*h,e*) for the probability of *h* given *e*, where we assume that P(*h,e*) is a probability function, that is to say that it satisfies the standard axioms of probability. It is one of the basic assumptions of Bayesianism that

$$C(h, e) = P(h, e)$$

or, in words that corroboration is a probability function. Popper denies this assumption, and claims instead that

$$C(h, e) \neq P(h, e)$$

or that corroboration is not a probability function.

Popper did not, however, divorce corroboration completely from probability. Although he thought that C (*h,e*) ≠ P(*h,e*), he nonetheless held that C(*h,e*) can be defined in terms of probabilities. In *LSD*, *ix, pp. 387–419, Popper gave various formulas for defining C(*h,e*) in terms of probabilities (see particularly pp. 400–401), but remarked that they can be considered as normalizations of the expression P (*e,h*) − P(*e*). This expression is thus central to Popper's account of corroboration.

So although Popper denied that corroboration was a probability function, he nonetheless held that it was a function of probabilities. This position raises the question of what interpretation is to be given to the probabilities used to define corroboration. It is important to realize that Popper did not adopt the propensity interpretation for such probabilities, but rather advocated a logical interpretation. Popper's logical interpretation differed from those of Keynes, Jeffreys, and Carnap in that Popper did not identify the logical probability of h given e with the degree of rational belief which should be accorded to h given e. As Popper himself puts it:

> while it is a mistake to think that probability may be interpreted as a measure of the rationality of our beliefs..., degree of corroboration may be so interpreted. As to the calculus of probability, it has a very large number of different interpretations. Although 'degree of rational belief' is not among them, there is a logical interpretation which takes probability as a generalization of deducibility. (*LSD*, pp. 414–415)

Popper has quite a number of arguments for his central claim that corroboration is not a probability function $(C(h,e) \neq P(h,e))$. I will here consider only one, the one which is, in my opinion, the strongest. This hinges on the important question of what probability should be assigned to universal laws, that is to laws of the form $(\forall x)F(x)$, where the quantifier ranges over a potentially infinite set of objects. A simple example of such a law would be: 'All ravens are black.' Now Popper claims that the prior probability $P(h)$ of any such law (h) is zero, but, Bayes's theorem states that

$$P(h, e) = P(e, h)P(h)/P(e)$$

So, if $P(h) = 0$, it follows that, for any e, $P(h,e) = 0$. Thus, if we accept that $C(h,e) = P(h,e)$, $C(h,e) = 0$, for any e. That is to say, universal laws will always have zero corroboration, no matter what evidence there is in favour. This Popper regards as absurd. He takes it as obvious that universal laws, which occur everywhere in science, can receive positive corroboration, and so rejects the Bayesian assumption that $C(h,e) = P(h,e)$. This is now Popper himself puts the matter: *'we may learn from experience more and more about universal laws without ever increasing their probability; . . .* we may test and corroborate some of them better and better, thereby increasing their *degree of corroboration* without altering their probability whose value remains zero' ('PIP', 383).

    This argument depends crucially on the claim that $P(h) = 0$ for any universal law h. Popper, who accepted the logical interpretation of probability, had a series of arguments designed to establish $P(h) = 0$ from this point of view. However these arguments came under fire from Colin Howson who, in his article 'Must the Logical Probability of Laws be Zero?'[15] attempted to show that all these arguments of Popper's are invalid, and that it is perfectly consistent to assign non-zero logical probabilities to universal laws. Howson's criticisms of Popper's arguments are very technical in character, and I will not attempt to summarize them here. Suffice it to say that Howson's critique seems to me entirely valid, and I agree with his conclusion that, if we do allow the existence of logical probabilities, then it is perfectly possible for the logical probability of a universal law to be non-zero.

    Having conceded this point to Howson, does this mean that I should abandon Popper's anti-Bayesian claim that corroboration is not a probability function ($C(h,e) \neq P(h,e)$)? In fact I still think that this claim is correct, but

[15] *British Journal for the Philosophy of Science*, **24** (1973), pp. 153–163.

obviously I have to alter Popper's original argument for the claim. In the remainder of this section I will propose such an alteration, and argue that it provides a new and stronger ground for holding that $P(h) = 0$, for any universal law h, so that, with its premise established in a different way, Popper's main argument for his claim that $C(h,e) \neq P(h,e)$ goes through. Unfortunately I have to add that my suggested alteration would, almost certainly, have been unacceptable to Popper himself. This is because it involves making some use of the subjective interpretation of probability, and Popper always opposed this approach to probability in the strongest possible terms. Thus he says, for example, 'These remarks should not be taken to mean that I am prepared to accept any form of the subjective interpretation;...' (*LSD*, p. 407). Thus in order to defend Popper's main thesis against Howson's attack, I am forced to use methods of which Popper would almost certainly have disapproved.

My suggestion then is that we should no longer interpret the probabilities which Popper uses to define degree of corroboration as logical probabilities. Indeed there are reasons, quite independent of Howson's critique, for making this move. There are many telling objections to the logical interpretation of probability, and it is doubtful whether such an interpretation of the probability calculus is really possible. The most famous critic of the logical interpretation of probability was Ramsey, who directed his attack specifically against Keynes's version of the theory. Keynes had claimed that our knowledge of logical probabilities is based ultimately on some kind of logical intuition which enables us to perceive these probability relations, at least in some cases. Ramsey criticized this appeal to intuition as follows:

> But let us now return to a more fundamental criticism of Mr Keynes' views, which is the obvious one that there really do not seem to be any such things as the probability relations he describes. He supposes that, at any rate in certain cases, they can be perceived; but speaking for myself I feel confident that this is not true. I do not perceive them, and if I am to be persuaded that they exist it must be by argument; moreover I shrewdly suspect that others do not perceive them either, because they are able to come to so very little agreement as to which of them relates any two given propositions.[16]

This is an interesting example of an argument whose validity seems to depend upon who puts it forward. Keynes had already remarked in his *Treatise on Probability* that: 'Some men ... may have a greater power of logical intuition than others.'[17] Thus, if a lesser mortal had put forward Ramsey's argument, Keynes might well have replied: 'Unfortunately it seems that you are one of those with a lesser rather than greater power of logical intuition.' However, as Ramsey was one of the leading logicians of his day, this retort was obviously not possible.

A second difficulty with the logical interpretation of probability is that it depends on what Keynes called the

---

[16] F. P. Ramsey, *Truth and Probability* (1926), in H. E. Kyburg and H. E. Smokier (eds), *Studies in Subjective Probability* (John Wiley), pp. 63–92 at pp. 65–66.

[17] P. 18.

Principle of Indifference to obtain numerical probabilities, and this principle gives rise to a whole series of paradoxes. Keynes is very frank about this, and chapter IV of his *Treatise on Probability* which deals with the principle of indifference contains what is perhaps the most complete account in the literature of the contradictions to which that principle leads.[18] Naturally Keynes tries to resolve these contradictions, but he is not very successful, and Ramsey aptly remarks: 'To be able to turn the Principle of Indifference out of formal logic is a great advantage; for it is fairly clearly impossible to lay down purely logical conditions for its validity, as is attempted by Mr Keynes.'[19]

We can see from this that there are many difficulties with the logical interpretation of probability, and it seems therefore worth considering the alternative theory proposed by Ramsey in England, and, independently by De Finetti in Italy.[20] This is the subjective interpretation of probability.

The starting point of the subjective theory is to identify probability with the degree of belief of a particular individual (Mr A say). Mr A's degree of belief in some event (E say) is then measured by seeing at what rate Mr A would bet on E under specified circumstances. Mr A's betting quotients are said to be *coherent* if his opponent in the betting (Ms B say) cannot make a *Dutch book* against him. Ms B makes a Dutch book against Mr A if she can succeed in arranging the bets so that she wins whatever happens. In the

[18] Pp. 41–64.   [19] Ramsey, *Truth and Probability*, p. 85.
[20] Ramsey, *Truth and Probability*; B de Finetti, 'Foresight: Its Logical Laws, Its Subjective Sources', in Kyburg and Smokier, pp. 95–158.

subjective theory the axioms of probability are obtained by what is known as the Dutch book argument, which leads to the Ramsey–De Finetti theorem. This states that Mr A's betting quotients are coherent if and only if they satisfy the axioms of probability.

The key point to note here is that the use of the Dutch book argument justifies the axioms of probability. This provides the foundation for the interpretation of probabilities as the subjective degrees of belief of individuals. However, it can also be used to give a foundation for other somewhat less subjective interpretations of probability. One such was adopted by Carnap in his later period, and also by Mary Hesse.[21] I have called it the 'topping-up' version of the logical interpretation of probability. The idea is to start with purely subjective degrees of belief. We then add one rationality constraint (coherence) to obtain the axioms of probability. However, if this can be 'topped-up' by further rationality constraints derived from logical or inductive intuition, we might get closer to a single rational degree of belief.

I have suggested another extension of the Dutch book argument.[22] The idea here is to consider not the degree of belief of a single individual (Mr A say), but the consensus degree of belief of a group of individuals A say, where

---

[21] Carnap, *Logical Foundations of Probability*, pp. 165–167; Mary Hesse, *The Structure of Scientific Inference* (Maxmillan, 1974).

[22] See D. A. Gillies, 'Intersubjective Probability and Confirmation Theory', *British Journal for the Philosophy of Science*, **42** (1991), pp. 513–533.

A consists of the individuals A_1, A_2, ..., A_n. It can be shown that if Ms B is betting not against a single individual Mr A, but against a group A of individuals, then she will be able to win money off the group as a whole, whatever happens, unless every member of the group agrees to adopt the same betting quotient. This common betting quotient can be taken as representing the consensus degree of belief of the group, and I call such a betting quotient, if it exists, the *intersubjective probability* of the group. What matters when we are considering the corroboration of a scientific theory by evidence is not so much the beliefs of individual scientists, as the consensus belief of the majority of experts working in the field. Thus for the theory of corroboration, the intersubjective interpretation of probability seems to me the most appropriate one.

For the purposes of the present argument, however, it does not matter whether the probabilities involved in the theory of corroboration are interpreted as subjective, as logical in the 'topping-up' sense of Hesse and the later Carnap, or as intersubjective. All these three interpretations identify probabilities in some way with coherent betting quotients, and use the Dutch book argument to obtain the axioms of probability. I will next argue that for any probability obtained by this betting approach, we must have $P(h) = 0$, for any universal law h.

The argument is indeed a very simple one, and has been put forward by a number of authors.[23] Take, for

---

[23] See L. J. Cohen, *The Implications of Induction* (Methuen, 1970), pp. 129–130.

example, h = All ravens are black, and suppose that Mr A is forced to bet on whether h is true. Mr A can never win the bet, since it can never be established with certainty that all ravens are black. However, Mr A might lose the bet if a non-black raven happens to be observed. Thus the only reasonable betting quotient for Mr A to adopt is zero. To put the point another way, if Mr A adopts any non-zero betting quotient, however small, then he might lose money, but can never win any. This shows that, if we introduce probabilities as betting quotients, we must have $P(h) = 0$ for any universal law h. Moreover, once this premise is established, the rest of Popper's argument given earlier goes through, and we can conclude that corroboration is not a probability function.

My conclusion is thus a somewhat ironical one. I think that one of Popper's main results in the theory of corroboration can be established, but only by using methods (the identification of probabilities with betting quotients, and the use of the Dutch book argument) which Popper would have abhorred.

6   Propensities and Indeterminism

DAVID MILLER

Prefatory Remarks

In these prefatory remarks, which are designed to locate my topic within the complex and wide-stretching field of Popper's thought and writings, I shall not say anything that those familiar with his work will not already know. Moreover, what I do say will take as understood many of the problems and theories, not to mention the terminology, that I shall later be doing my best to make understandable. My apologies are therefore due equally to those who know something about Popper's discussions of indeterminism and of the propensity interpretation of probability, and to those who know nothing.

The Postscript to The Logic of Scientific Discovery was eventually published in three volumes (RAS, OU, QTSP) in 1982 and 1983. It had been conceived in the first half of the 1950s as a series of new addenda to the parent work, but had grown at such a pace that the decision was eventually taken to separate it from its source and prepare it for independent existence. The Logic of Scientific Discovery (LSD), the English translation of the original Logik der Forschung (LdF) of 1934, made its own way, appearing in 1959 still burdened with new addenda (twelve of them in the English edition, increasing to no fewer than twenty in the 10th German edition of 1994). But *The Postscript* did not follow, and was eventually withdrawn at galley-proof stage, to be subjected to a strenuous process of rewriting. As so often happens, new urgencies emerged, and by the early 1960s the project was as good as shelved. Only in about 1980 did the late Bill Bartley—who had been considerably responsible for the postponement of the book's parturition—undertake to prepare it for publication. And so, like its precursor, it first appeared in English some twenty-five years after its origination, and like its precursor heavily encumbered with additional material.

When *LSD* was published in 1959, the anonymous reviewer in *The Times Literary Supplement* described it as a 'remarkable book', and declared: 'One cannot help feeling that if it had been translated as soon as it was originally published philosophy in this country might have been saved some detours.' A strikingly similar remark is made by Magee in the preamble to one of his published conversations with Popper.[1] I doubt that the same is true of *The Postscript*, at least to the same degree, since a fair number of its central themes were to appear in Popper's writings throughout the 1960s and 1970s,[2] often developed substantially beyond their *Postscript* versions, and others seem to have been introduced into public discussion by colleagues, several of

---

[1] Bryan Magee, *Modern British Philosophy* (London: Secker and Warburg, 1971), p. 66.

[2] Compare the assessment of *The Postscript* on p. 455 of Anthony O'Hear's critical notice, *Mind* **94** (1985) pp. 453–471.

whom (including myself for one weekend) had had access to the proofs. Nonetheless, it is unfortunate that the book appeared as late as it did. By the 1980s, some parts of it, especially parts of Volume I, *RAS*, seemed dated, fighting battles long since won (either convincingly or by default) against opponents long since retired. The propensity interpretation of probability, the thread running through all three volumes, had been appropriated and articulated by others, and—if I may say so—perhaps misunderstood by some. Most seriously, some of the book's most challenging ideas found themselves overshadowed by independent developments. I am thinking here especially of the attack in Volume II, *OU*, on scientific determinism, which anticipated (but came after) the current fascination with non-linear dynamics. In this last case the situation looks at first glance to be a somewhat peculiar one, since the standard view is that the theory of chaos lends credibility to determinism, whereas Popper used very similar ideas to cast doubt on determinism. Although the situation is not as paradoxical as that makes it sound, it does deserve to be looked at, and to be resolved.

The two topics of my title, propensities and indeterminism, are indeed closely interconnected in the original version of *The Postscript*. The propensity interpretation of probability, it is made clear there, makes limited sense—more exactly, it has limited interest—in a world strictly controlled by deterministic laws. Moreover, one of the central arguments used against metaphysical determinism (the argument known as Landé's blade) also illustrates the need for an objectivist interpretation of probability deeper than the frequency interpretation, and the propensity interpretation certainly has some title to be able to fill that role. (No matter that many, myself included, doubt whether Landé's blade is a sound argument. For if it fails as a criticism of determinism, it fails also as an illustration of the potency and usefulness of the propensity interpretation.) But it is not only in *The Postscript* as originally conceived that there is a close and intimate liaison between propensities and indeterminacies. In later papers, especially the 1965 lecture 'Of Clouds and Clocks' ('OCC', reprinted in *OK*), and the aptly entitled addendum 'Indeterminism is not Enough' ('IINE'), which helps to pad out *OU* (Volume II of *The Postscript*), Popper maintained that for human freedom and creativity to flourish it is not enough that we live in an indeterministic physical universe. In addition, the physical universe must be causally open to influences from outside, not only mental activity but also abstract influences such as arguments and critical discussions. Even later, especially in the lecture *AWP*, he came to stress the idea that, just as the propensity interpretation is meagre fare without indeterminism, so indeterminism is meagre fare without full-blooded propensities. In section 4, I hope to be able to provide some flavour of the enriched diet of propensities + indeterminism.

No single paper could deal adequately with the two topics of my title, and I shall make no more than incidental references to the work of other philosophers. Even the entirety of Popper's own writings against determinism is beyond my power to survey. But it would be wilfully negligent to disregard altogether the central arguments of *PH*, to the effect that there are no laws of development (let alone progress) in history, and that no scientific pre-diction of the course of human history is conceivable. Let me mention briefly the most celebrated of these arguments: the argument that the growth of scientific knowledge is unpredictable (*OU*, section 21), and that since the course of human history depends decisively on the way in which scientific knowledge grows, human history is unpredictable too (the Preface to *PH*).

For all its popularity, this two-stage argument, it seems to me, is not quite satisfactory (which is not to say that its conclusion is not correct).

In the first place, as Popper admits, its intermediate conclusion may be false; it may indeed be the case (as some contemporary physicists appear to think) that 'the growth of our theoretical knowledge has come to an end' and that all that is left is 'the endless task of applying our theories to ever new and ever different initial conditions' (*OU*, p. 67). Moreover, our inability to advance further theoretically may itself be one of those things that our present theories allow us to predict. That is, we may be able to predict from what we know now that the future growth of theoretical knowledge will be nil. More seriously, however, the second stage of the argument is invalid. For the unpredictability of a cause does not always mean the unpredictability of its effect. (All deaths are more or less predictable, though their causes often are not.) Science fiction demonstrates that uncannily accurate (if not very precise) predictions of the advance of applied science, especially technology, need not depend on any prediction of the appearance of any new scientific theory. (Science fiction is indeed mostly technology fiction, and genuine science fiction would be just a kind of—perhaps too easily falsifiable—science.) Against this it may be claimed that no scientific, rather than clairvoyant, prediction of an effect is possible without knowledge of its cause; to which the only correct retort is that technological advance is not in this sense an effect of scientific advance. It is natural to suppose that, if this is so, it is because much new technology is no more than an application of already existing science, but this response misses the point that what scientific theories tell us is what cannot be achieved, not what can be (*PH*, sections 20, 26), so that new theories alone cannot be the source of new technology.[3] In other words, even if we were to predict correctly the future of theoretical science, we should hardly begin to be able to predict its impact. Much more important for any rational prediction of the future of technology is knowledge of what needs to be accomplished. Having said this, I at once acknowledge that new theories may bring to our attention previously unknown phenomena —such as the untapped energy resources within the atom— that can be exploited if we are inventive enough. It would be foolish to claim that the advance of theoretical science has no significance for the advance of technology.

I want to stress only that it is not the primary source, and that the applications of science may be more predictable than science is itself.

[3] The point is elaborated in Chapter 2.2g, pp. 39–41, of David Miller, *Critical Rationalism. A Restatement and Defence* (Chicago and La Salle: Open Court, 1994).

To be sure, such predictions (like nearly all predictions in social life) will be rather imprecise, predictions in principle rather than predictions in detail. But the dogma of historicism that Popper attacked in *PH* was likewise concerned with the broad sweep of human history, rather than with its intimate details. Determinism is different. In its most provocative form it holds that the future of the world is predictable, at least in principle, in uncompromising detail. To this doctrine we now turn.

1. Determinism and Deterministic Theories

Determinism and indeterminism, which are mutual contradictories, may be amongst those philosophical theories— another is realism—that it is much easier to have an opinion on than to have a decent formulation of. Popper begins by listing a variety of deterministic positions: religious, scientific, metaphysical determinism. The last of these is characterized (*OU*, p. 8) as the view that

> all events in this world are fixed, or unalterable, or
> predetermined. It does not assert that they are known to
> anybody, or predictable by scientific means. But it asserts
> that the future is as little changeable as is the past.
> Everybody knows what we mean when we say that the
> past cannot be changed. It is in precisely the same sense
> that the future cannot be changed, according to
> metaphysical determinism.

In many respects this is a clear statement. In fact, if it were to be presented together with an idealistic theory of time, according to which the future is just as real or unreal as is the past, it would be unexceptionable. The trouble is that it is simply unbelievable as it stands. It appeals to the common-sense view that the past is out of bounds, yet it is yoked to a theory that is starkly counter to common sense. Common sense undoubtedly regards time as a real process, and maintains that the future, even if determined, does not possess the same kind of established (if inaccessible) reality that the past possesses. There is a difference between past and future that is incontrovertibly relevant to the claim that the past cannot be changed; and that is that the past has happened and the future has not. Russell suggested that 'but for the accident that memory works backward and not forward, we should regard the future as equally determined by the fact that it will happen',[4] but the kind of logical determination at issue here is, as he conceded,[5] less than the whole story. The image of the history of the universe as already developed on a cinematographic film, in short, the idea of a block-universe, is indeed a common concomitant of metaphysical determinism. Still, films pass through projectors, and this passage is a real process even if everything to be seen on the screen is fixed before the process begins. I should have thought that it should be possible to be a determinist yet to hold that the future is not yet realized. (See also *OU*, sections 11, 18, and 26.)

---

[4] Bertrand Russell, 'On the Notion of Cause', *Mysticism and Logic* (London: Allen & Unwin, 1918), p. 202; (Melbourne, London, & Baltimore: Pelican edition, 1953), p. 190.

[5] Ibid., p. 203; Pelican edition, pp. 191f.

Be that as it may, the doctrine of metaphysical determinism—which Popper says 'may be described as containing only what is common to the various deterministic theories' (*OU*, p. 8)—is plainly a metaphysical doctrine in the sense that it is not open to empirical falsification. No observational or experimental results, however little anticipated, can demonstrate that what is observed to take place was not fixed prior to the observation or the experiment. Nor can they demonstrate the opposite. Neither metaphysical determinism nor metaphysical indeterminism (the doctrine that some events, perhaps not many, are not fixed), that is to say, is open to direct scientific investigation. Now it is certainly no part of Popper's position in *OU* (as it perhaps was in *LdF*) to restrict rational discussion or investigation to the theories of empirical science—and perhaps also those of logic and mathematics. The determinism/indeterminism debate may be seen as a salient test case for the contrapositivist view that metaphysical theories, if formulated clearly, can be rationally discussed and criticized; that the modes of rational discussion are not limited to empirical research and logical and mathematical analysis. It is this concern that leads Popper to the formulation of what he calls 'scientific determinism' (the topic of the next section). Let us see how that emerges.

One way in which metaphysical theories may be discussed is in terms of the problems that they solve, or were designed to solve but fail to solve (*CR*, Chapter 8). This works well for Greek atomism, to take a shining example, but it seems much less effective for determinism and indeterminism. For these doctrines seem more like attempts to evade problems than genuine attempts at solutions. Determinism does little more than generalize from the undoubted truth that many aspects of our lives evince vivid regularities and unexpected repetitions, regularities that it is not in our power to alter; the residual problem, simply deferred or legislated away, is how to account for the irregularities that permeate the regularity, for the endless novelty whose irregular but not infrequent appearance we almost take for granted, and for those regularities that we do seem to be able to alter and even to insist on. With indeterminism it is partly the other way round: the clouds are comprehensible, but not the clocks. It is part of Popper's thesis in *OU* that there is indeed an asymmetry here; or rather, that there are two asymmetries. In the first place, indeterminism does not suggest that no events are fixed in advance, only that there are some that are not (*OU*, p. 28). As Earman points out, this irenic doctrine may not be as innocuous as it seems,[6] which does not imply that indeterminists are not free to espouse some kind of partial determinism (*OU*, pp. 126f.), and to explain clocks as more or less deterministic systems more or less isolated from external influence. Secondly, Popper claims (*OU*, section 28), indeterminism (coupled with the propensity interpretation of probability) has some ability to explain statistical regularity, whereas determinism is impotent to explain irregularity (rather than just postulate it as an accumulation of unrelated effects). Were this latter claim true, there might be a case for

[6] John Earman, *A Primer on Determinism, The University of Western Ontario Series in Philosophy of Science*, volume 32 (Dordrecht: D. Reidel Publishing Company, 1986), pp. 13f.

supposing that indeterminism had a problem-solving edge. But nothing that I know at present gives me much optimism that it is true.

There is another way, almost a scientific way, in which metaphysical theories may be evaluated, and that is by revealing them as logical consequences of scientific theories, hence true if those theories are true. If suitable scientific theories are not available, they may have to be invented; it was by this process that the bald existential hypothesis 'There exists neutrinos' eventually became incorporated into science. Is there then any way in which either metaphysical determinism or metaphysical indeterminism could be distilled from any scientific theory?

Once we pose the question in this form, we see that the matter is not straightforward. These doctrines, as formulated above, are concerned with the fixedness or unfixedness of the future, with what is or is not alterable. Scientific theories are not formulated in this kind of vocabulary.[7] This helps to explain how it is that determinism and indeterminism are both irrefutable. It cannot be true that each is, as Popper says of determinism, 'irrefutable just because of its weakness' (*OU*, p. 8), since the contradictory of a weak theory is correspondingly strong; indeed, it is the relative weakness of indeterminism that he usually stresses elsewhere (*OU*, p. 28). It might be better to say that determinism, being universal, is a strong theory, but nonetheless an irrefutable

---

[7] Compare Russell's remark that 'in advanced sciences such as gravitational astronomy, the word "cause" never occurs' (Russell, 'On the Notion of Cause', p. 189; Pelican edition, p. 171).

one because not properly concerned with matters to which empirical investigation has direct access.

No scientific theory, as normally understood, logically implies metaphysical determinism in the formulation above, for no theory says that what it says will happen has to happen. Now perhaps the fault here is with the way that science usually formulates its theories, paying little attention to its own metaphysical heritage and ruthlessly eliminating metaphysical and modal elements. But even those who emphasize the metaphysical content of science distinguish between substantial metaphysics and idle metaphysics, metaphysics that does not contribute to testability or empirical content.[8] Yet it does not seem that by rewriting a scientific theory so that its predictions come garnished with necessity we make it magically more testable. The only way of showing that something is not bound to occur is to show that it does not occur.

Classical celestial mechanics is presented by Popper as a prime example of what he calls a *prima facie* deterministic theory; that is to say (*OU*, p. 31), it is a theory that

> allows us to deduce, from a *mathematically exact* description of the initial state of a closed system which is described in terms of the theory, the description, *with any stipulated finite degree of precision*, of the state of the system at any given future instant of time.

(Complete precision cannot be expected, because some families of equations are soluble only by approximation

---

[8] John Watkins, *Science and Scepticism* (London: Hutchinson 1984), p. 205.

methods.) But even if classical mechanics satisfies this condition, which is a more controversial claim than Popper seems to have appreciated,[9] it does not imply metaphysical determinism. If **N** is classical mechanics, *I* an appropriate package of auxiliary hypotheses and initial conditions relevant to some isolated physical system, and *f* a description of some future state of the system, then of course **N** ∧ *I* logically or necessarily implies *f*. But for familiar reasons it may be concluded that *f* is unalterable or necessary only if it is assumed that both **N** and *I* are unalterable. We may be quite prepared to make these assumptions, especially with respect to initial conditions *I* that relate entirely to events in the unalterable past, but it must be recognized that they are additional assumptions and not parts of classical mechanics.

It may be felt that these considerations give too much weight to the emphasis on unalterability that appears in the intuitive statement of determinism, and that it is enough to call the world deterministic if its behaviour is entirely described by a *prima facie* deterministic theory together with statements of initial conditions. There are well known perils in this approach, which we escape only if we are careful not to count any deductively closed accumulation of sentences as a theory.[10] If we suppose that we can do this —it seems to require some distinction between lawlike statements and others, and therefore some concession to the demand that the future of a deterministic world is bound

[9] Earman, *A Primer on Determinism*, Chapter III.

[10] Russell, 'On the Notion of Cause', pp. 203–205, Pelican edition, pp. 192–194, and Earman, *A Primer on Determinism*, Chapter II.5.

by law, unalterable—we may be tempted to put meat on Popper's suggestion that metaphysical determinism contains 'only what is common to the various deterministic theories' (*OU*, p. 8). But it is by no means obvious that the outcome (the intersection $\vee \mathcal{D}$ of the class $\mathcal{D}$ of all *prima facie* deterministic theories) would be a recognisable formulation of determinism. Although it is trivial that *prima facie* deterministic theories have something in common, it is not obvious that they share any non-trivial logical consequences. The failure is plain in the complementary case of indeterminism. Since many *prima facie* indeterministic theories, if not all, have *prima facie* deterministic extensions, what is common to them all, the intersection $\vee I$ follows from many *prima facie* deterministic theories too, and can therefore scarcely be a formulation of indeterminism. (Indeed, it seems unavoidable that $\vee I = \mathbf{T}$, the class of logical truths.) What all this indicates is that there is no available statement of metaphysical determinism that does not make explicit reference to laws ($\vee \mathcal{D}$: might provide such a statement even though it is defined in terms of laws, but sadly we do not know what theory $\vee \mathcal{D}$ is). The only formulation of metaphysical determinism that is implied by every *prima facie* deterministic theory seems to be the metalinguistic thesis that some *prima facie* deterministic theory is true. Metaphysical indeterminism must be stated as the negation of this thesis (rather than the thesis that some *prima facie* indeterministic theory is true).[11]

---

[11] Compare Earman, ibid., p. 13, who limits himself to distinguishing between deterministic and indeterministic worlds. As far as I can see,

2. Scientific Determinism

By such means the doctrine of metaphysical determinism can be formally incorporated into science, provided that science contains a theory that is, as classical mechanics was supposed to be, *prima facie* deterministic. But it is plain that its position is peripheral and precarious, and that only if the theory is strictly true and quite comprehensive is anything said about whether determinism is true. This might not matter except that the comprehensiveness even of Newtonian physics has never been more than a dream (*OU*, p. 38), and—given the pervasiveness of errors and approximations—the same might be said, and was said by Newton himself and by Peirce (*OK*, pp. 212f.), of its claim to be strictly true.[12] It would therefore be valuable to be able to formulate a deterministic position that digs somewhat deeper into science, and does not depend so crucially upon the existence of a virtually perfect theory.

A considerable part of *OU* is concerned with a doctrine labelled 'scientific determinism'. Popper (*OU*,

---

the only formulation of determinism given is that the actual world is a deterministic world. Much of Earman's book is devoted to the task of investigating whether the principal theories of classical and modern physics are *prima facie* deterministic.

[12] Nancy Cartwright, *How the Laws of Physics Lie* (Oxford: Clarendon Press, 1983), and Macolm Forster and Elliott Sober, 'How to tell when Simpler, More Unified, or Less *Ad Hoc* Theories will Provide More Accurate Predictions', *The British Journal for the Philosophy of Science* 45 (1994), pp. 1–35, are examples of contemporary writers who, not always for identical reasons, reject the view that real science is exact science.

p. 30) attributes the substance of this doctrine to Laplace,[13] who famously imagined

> a superhuman intelligence, capable of ascertaining the complete set of initial conditions of the world system at any one instant of time. With the help of these initial conditions and the laws of nature, i.e., the equations of mechanics, the demon would be able, according to Laplace, to deduce all future states of the world system...

Scientific determinism is this idea of predictive potency cast in a testable form. Popper seems to intend it to be a scientific doctrine in at least three distinguishable respects:

(i) like metaphysical determinism in its final formulation above (but not in its initial intuitive formulation), it concerns science, and makes explicit reference to scientific theories;

(ii) it restricts itself to real scientific capacities, in particular predictive capacities;

(iii) unlike metaphysical determinism, it is itself open to direct scientific investigation; that is, it is empirically falsifiable.

Let us take these points in turn.

(i) Like metaphysical determinism, scientific determinism is, for all its reference to scientific laws, a doctrine about the world, not about our knowledge of the world. Popper writes, for example, that 'the fundamental

---

[13] P. S. Laplace, *Essai philosophique sur les probabilites* (1819).

idea ... is that the structure of the world is such that every future event can in principle be rationally calculated in advance, if only we know the laws of nature, and the present or past state of the world' (*OU*, p. 6). A little later he declares unequivocally that '[i]n asserting [scientific determinism], ... we assert *of the world* that it has a certain property' (*OU*, p. 38).

(ii) Scientific determinism is stronger than metaphysical determinism, and says more than that the world is ruled by a *prima facie* deterministic theory (*OU*, section 13), or that 'every future event can in principle be rationally calculated in advance...'. That would mean only that a completely precise description of the state of a system at an initial time could be transformed formally into a description (at any demanded level of precision) of its state at a later time. Scientific determinism says in addition that the formal mathematical relation in the theory that subsists between the earlier description and the later one can in principle be realized in a physical process (not merely in a mathematical calculation); the process begins with the collection of information appropriate to the formulation of the description of the system at the earlier time, and terminates in the formulation and publication of a prediction prior to the later time. Since this process is a physical one, the initial information cannot be expected in a perfectly precise form, though no finite bound to its precision need be insisted on. In short, the predictability that scientific determinism adds to metaphysical determinism is supposed to be what Popper calls 'predictability from within' (*OU*, section 11). It must be feasible, not just possible, and must make use of real physical processes.

(iii) Scientific determinism will not itself be a scientific doctrine if it says only that, given sufficiently precise initial conditions, a prediction may be obtained at any pre-assigned level of precision; for failure to achieve a satisfactory prediction could always be blamed on an insufficiently precise starting point. It is therefore necessary to require that the *prima facie* deterministic theory with which we have equipped ourselves should satisfy also what Popper calls the principle of accountability (*OU*, pp. 12f., emphasis suppressed). In its most appropriate form this says that, given any prediction task

> we can calculate from our prediction task ... the requisite degree of precision of ... the results of possible measurements from which the initial conditions can be calculated.

That is, for any level of precision demanded for the prediction, we may work out in advance, with assistance from the theory, how precise we need to make our measurements of those quantities from which we calculate the initial conditions.

Popper (*OU*, pp. 36f) offers two versions of scientific determinism satisfying the requirements (i)–(iii)

(I) the state of any closed physical system at any given future instant of time can be predicted, even from within the system, with any specified degree of precision, by deducing the prediction from theories, in conjunction with initial conditions whose required degree of precision can always be calculated (in accordance with the principle of accountability) if the prediction task is given.

(II) whether or not a closed physical system will at any given future instant of time be in any given state can be predicted, even from within the system, by deduction from theories, in conjunction with initial conditions whose required degree of precision can always be calculated (in accordance with the principle of accountability) if the prediction task is given.

The stronger version (II) is of historical significance, since it subsumes the problem, by which Laplace was exercised, of whether classical mechanics ensures that the solar system is dynamically stable. It is the version of scientific determinism that is refuted by considerations from non-linear dynamics.

A number of things could be said about the details of these formulations of scientific determinism. Here it suffices to note that not only the precision of predictions, but also their accuracy [closeness to the truth], needs to be mentioned if scientific determinism is to be a statement about the world, as intended in (i), rather than a statement about our theories about the world. I shall take this amendment for granted. About Laplace's demon, which he takes to be a forerunner of scientific determinism, Popper (ibid., p. 30) writes:

> The crucial point about this argument of Laplace's is this. *It makes the doctrine of determinism a truth of science*

rather than of religion. Laplace's demon is not an omniscient God, merely a super-scientist.

and a few pages later (ibid., p. 33):

The crucial point is that ... [scientific determinism] appeals to *the success of human science*, such as Newtonian theory: 'scientific' determinism is to appear as a result of the success of empirical science, or at least as supported by it. It appears to be based upon *human experience.*

It is perhaps not wholly clear that in the formulations given above scientific determinism is a scientific theory in its own right, as intended by the manufacturer. For could not any failure to make accurate predictions of sufficient precision, once the required precision of the initial conditions has been assured, be attributed not to the failure of scientific determinism itself, but to the falsity of the attendant explanatory scientific theory (say, classical mechanics)? Although the answer to this question is of course affirmative, two points should be made. The first is that we ought to regard an explanatory theory as falsified only if we obtain a reproducible effect that contradicts it (*LSD*, section 22). But a possibility is that by sharpening the initial conditions we obtain predictions of varying precision (and indeed, accuracy), in which event we would do well not to reject the explanatory theory, but to reject the deterministic suggestion that better outputs can always be obtained from better inputs. I shall return to this point in a moment. In the second place, determinism is not a doctrine merely about the possibility of calculating what initial measurements are required for the delivery of particular predictions; as noted in (ii), it says also that these measurements are physically obtainable and the predictions physically deliverable. It was for this reason that I followed Popper above in stating the principle of accountability in terms of measurements rather than initial conditions; indeed, he says that '"scientific" determinism requires accountability in the stronger sense ... a theory ... which is non-accountable in the strong sense would be one whose determinist character could in principle not be tested by us ... it could not be used to support "scientific" determinism' (OU, p. 13).[14] Recurrent failure sufficiently to refine measurement techniques or speed up prediction techniques would therefore count against scientific determinism unless, of course, the bare possibility of unknown techniques was appealed to. That would, it must be admitted, undermine (and perhaps even destroy) any claim to empirical status.

So understood scientific determinism may seem to be not only falsifiable but plainly false. It requires the possibility of measurements made at unbounded levels of precision, and—what is worse—computations and publications generated at unbounded speeds. Predictions of any complexity are, remember, required to be possible for any future

---

[14] The distinction is not registered by Honderich, who concludes that 'successful predictions within neuroscience *are* evidence for determinism, even overwhelming evidence'. The version of determinism espoused by Honderich does not maintain that measurements of ever-increasing precision are possible—indeed, it seems to deny it—and accordingly it falls short of—or contradicts—scientific determinism. See Ted Honderich, *A Theory of Determinism* (Oxford: Clarendon Press, 1988), Chapter 6.6, especially p. 356.

instant. Scientific determinists can take little comfort in the thought that the closer the initial and final times are, the less laborious should the computation have to be; this may be so —it may not be—but in general the presentation of the results can be expected to consume more time than does their preparation. Doubtless scientific determinism could be modified to deal with this piece of trouble, without serious compromise either to its deterministic character or to its scientific status. After all, if publishable predictions for tomorrow are promised, we may cheerfully condone the practical inability of the predictor to tell us what to expect in the next picosecond. It is interesting that pre-dieting devices are usually more at a loss in the long term than in the short term. But if the dissemination of the prediction is an ineliminable part of the task, then short notice can be bad news too.

The main arguments against scientific determinism, however, are of a more logical or analytical kind than these. Popper argues in particular against the possibility of self-prediction, on the grounds that the activity of predicting (which is a physical activity) unremittingly encroaches on and disturbs the state of the predictor, whose future is what the predictor is aiming to predict (*OU*, sections 22f.). I am unable to evaluate this complex argument on the present occasion, but let me note that the interference of the predictor with its own future is no mere reiteration of the familiar point that every act of observation disturbs what is being observed (see *OU*, note 1 on p. 35). Nothing in scientific determinism requires a predictor to be able to predict how a system would have evolved had the initial measurements not been made on it. What is asked is only that sufficiently precise measurements can be performed to yield a prediction of the behaviour of the disturbed system. For this reason I am unconvinced that scientific determinism needs to be linked to a *prima facie* deterministic theory that is comprehensive, or even that it implies metaphysical determinism. Indeed, the very possibility of manipulating initial conditions at will suggests that scientific determinism presupposes metaphysical indeterminism.

Let me turn to something widely agreed, by determinists and indeterminists alike: the fact that if classical mechanics (and most other non-linear theories) are true then scientific determinism is false. Although very briefly treated, this is one of the principal theses of *OU*. It is one of the principal consequences too of the theory of dynamical chaos. The difficulty for scientific determinism is that in many interesting theories an accurate prediction of any precision at all can be achieved only by requiring unlimited precision in (the measurements of) the initial conditions. In a section of less than two pages (*OU*, pp. 39f.) Popper resurrects a result by Hadamard discussed penetratingly in 1906 by Duhem,[15] which makes just this point. Imagine projectiles being thrown on to an infinite surface. Then some may settle into closed orbits, while some disappear to infinity. Hadamard established that on some simple varieties of surface of negative curvature there is no way, short of absolute precision, in which the angle of launch can be arranged so as to ensure

[15] Pierre Duhem, *The Aim and Structure of Physical Theory* (Princeton University Press 1954), Part II, Chapter 111.3.

either that the projectile falls into a loop or that it flies off for ever. Between any two launching angles that lead to loops there are some that lead to infinity; and between any two launching angles that lead to infinity there are some that lead to loops. The stronger form (II) of scientific determinism stated above is thus false, even if classical physics is true. No matter how precisely the launch is managed (short of absolute precision), qualitatively different outcomes are possible.

A more familiar example is the logistic function

$$f(t+1) = \lambda f(t)(1 - f(t)),$$

where $f(0)$ lies in the open interval $(0,1)$ and λ is in $[0,4]$. Here too scientific determinism in the form (II) is contradicted, except for low values of λ and special values of $f(0)$.[16] For example, if $\lambda = 4$ and $f(0) = 0.75$ then f remains constant for ever. But almost every other initial value *off* leads eventually to an irregular, apparently random, curve. No amount of initial precision short of absolute precision will allow us to predict that f remains constant, though if $f(0) \neq 0.75$ we have only to discover that fact by a sufficiently delicate measurement in order to be able to predict that f will not remain constant. But (II) is infringed. On the other hand, the weaker version (I) of scientific determinism is not controverted, since for every i the value of $f(i)$ is a continuous function of $f(0)$.

---

[16] For details consult almost any book on chaos, for example, Ian Stewart, *Does God Play Dice?*, (Oxford: Blackwell, 1989), pp. 155–164. A graph of the first 400 values of the logistic function for $\lambda = 4$, $f(0) = 0.75000000012$ may be found in Miller, *Critical Rationalism*, p. 155.

Where do Hadamard's result, and the logistic function, and the impossibility of self-prediction leave scientific determinism? It is clear that a *prima facie* deterministic theory may be true, yet scientific determinism false. Not only do *prima facie* deterministic theories not imply scientific determinism, some of them imply the negation of scientific determinism. Nothing follows, as far as I can tell, about the possibility of a more local kind of determinism that equally 'appeals to the *success of human science*, ... [and] appears to be based upon *human experience*' (*OU*, p. 33), and concerns only predictability (up to a certain level of precision) in prepared systems isolated from the predicting apparatus.

Where do Hadamard's result and the logistic function, where does the attack on scientific determinism, leave metaphysical determinism? It is clear that whether or not it implies metaphysical determinism, the falsity of scientific determinism in classical physics tells us nothing about the truth or falsity of metaphysical determinism. Earman regards scientific determinism as 'such a wrong-headed conception of determinism' that its failure is without significance.[17] Others[18] seem ready to conclude that metaphysical determinism has been vindicated. Popper himself maintains that, because indeterminism is what we would now call the

[17] Earman, *A Primer on Determinism*, p. 9. It is on quite different grounds that Earman himself takes metaphysical determinism to fail in classical physics.

[18] For example, G. M. K. Hunt, 'Determinism, Predictability and Chaos', *Analysis* **47** (1987), pp. 129–133.

default position, 'the burden of proof rests upon the shoulders of the determinist' (*OU*, p. 27). But it would have been much better to have said that the burden of criticism, and of exposing one's position to criticism, rests on all parties to a dispute. By being prepared to replace the unfalsifiable doctrine of metaphysical determinism by a stronger scientific thesis, determinists certainly stuck their necks out (to employ a characteristic Popperian expression), and they lost the gamble. That does not show metaphysical determinism to be false, but it does serve to emphasize how exposed is its position within the body of scientific knowledge—far more exposed, I should have thought, than that of metaphysical theses such as the principle of conservation of energy,[19] whose regulative role is undeniable.[20] If our fundamental theories turn out not to be *prima facie* deterministic, then that is the end of metaphysical determinism. It is now almost a commonplace that this is what has happened with the advent of quantum mechanics. Science has shown metaphysical determinism to be false.

3. Landé's Blade

Rather than discuss quantum mechanics, which I have no qualifications to do, I want to look here at the central philosophical argument in *OU* against metaphysical determinism. It is an argument to the effect that there are

[19] Earman, *A Primer on Determinism*, p. 10.
[20] Emile Meyerson, *Identity and Reality* (London: Allen & Unwin, 1930), Chapter V.

phenomena that only indeterministic theories can explain satisfactorily.

One of the most persistent themes in Popper's writing on probability theory over sixty years has been the claim that statistical conclusions can be obtained only from statistical or (latterly) probabilistic premises, and not from deterministic premises. In Chapter VIII of *LSD*—a book more deterministic than indeterministic—he singled out two central problems for any theory of physical probability. One, the problem of the falsifiability of probability statements, though important, need not detain us here.[21] The other, the fundamental problem of the theory of chance, as he called it (*LSD*, section 49), is the problem of explaining the statistical stability witnessed in many otherwise disordered sequences encountered in science and in everyday life. How are we to explain the appearance, in the midst of disorder, of a very tightly controlled order? Sequences of tosses of a die on a well regulated piece of apparatus, of results on a roulette wheel, of incidents of accidents at a busy street intersection, of the appearance of various characteristics in successions of generations in genetic experiments, and so on, all show the most remarkable statistical stability. Not only is it possible to estimate relative frequencies of 6-UP, or of EVEN, or of head-on crashes, or of blue eyes, by considering the frequencies shown in finite sequences, but these relative frequencies characteristically settle down very

---

[21] It is the central topic of Part III of D. A. Gillies, *An Objective Theory of Probability* (London: Methuen, 1973). See also Miller, *Critical Rationalism*, Chapter 9.3.

rapidly near to their final values, and thereafter do not deviate much from those values. Provided the macroscopic conditions that generate the sequences are unchanged, the sequences of outcomes are found to be remarkably stable. 'The tendency of statistical averages to remain stable if the conditions remain stable is one of the most remarkable characteristics of our universe' Popper writes (*AWP*, p. 12). Here is a problem for determinism and indeterminism alike. For indeterminists—who may be tempted to regard such sequences as sequences of wholly undetermined events—it is the problem of explaining why any order at all should appear in the midst of disorder; truly disordered sequences might be expected to be just that, without sign of constancy. For determinists, the problem is rather one of understanding disorder at all; and, given that it does occur, of explaining its relatively constant features.

Traditionally indeterminism has been able to give an explanation of statistical stability by construing these undetermined events as events in the thrall of a probability distribution. In von Mises's version of the frequency interpretation of probability, to be sure, the explanation did not amount to a great deal. Since probabilities are defined only for sequences in which frequencies converge, statistical stability is not much more than another way of saying that the type of event in question has a probability of occurring. For this and other reasons Popper felt obliged in *LSD* to abandon von Mises's axiom of convergence, and to show that it could be deduced from an appropriately invigorated version of the axiom of randomness or excluded gambling systems. Even so, the explanation is hardly a very deep one, since randomness is explained in terms of the persistence of frequencies under various forms of selection of subsequences. There are frequencies mentioned in the premises, and it is hardly to be wondered at that there are frequencies mentioned also in the conclusion. The question of why it is frequencies that converge, that it is frequency-statistics that are stable, is not handled in depth in the frequency interpretation. One of the virtues of the propensity interpretation of probability is that it offers a somewhat deeper explanation of statistical stability.

Probability as such is not my topic in this paper, and I do not intend to describe the propensity interpretation of probability in much detail. Two of its principal features are that, unlike the frequency interpretation, it ascribes probabilities sensibly to single events, rather than to types of events, and that it relativizes probability ascriptions to the whole current state of the universe, rather than to the immediate locality. The probability of the occurrence of 6-UP WITH THIS DIE THROWN NOW is fixed in general not only by the physical features of the die, and by the features of the apparatus being used to throw it, and its immediate surroundings, but by the whole state of the world. The probability is said to be a measure of the propensity of the world to develop into one in which the outcome of the throw of the die is 6-UP. The propensity that is the probability is not a propensity of the die, or of the apparatus, but of the world. It is not, as in the frequency interpretation, relative to a reference class.[22] As so

[22] Hence the criticism of the propensity interpretation given in section 3.3 of Colin Howson, 'Theories of Probability', *The British Journal for*

understood, probabilities can take values other than 0 or 1 only if the world is metaphysically indeterministic—that is, in this case, the probability of 6-UP WITH THIS DIE THROWN NOW lies strictly between 0 and 1 only if the outcome of the throw of the die is not fixed in advance; the world has a propensity, which is neither cast-iron necessity nor cast-iron impossibility, to develop in the way described. If propensities, so construed, are postulated to satisfy the axioms of the calculus of probability, then we may conclude, through the laws of large numbers, that in stable circumstances, in which the propensities of the events in question do not change, there is an overwhelming propensity for a longish sequence of throws to be statistically stable. Now again this explanation is not a magically deep one, since the probability axioms are plainly satisfied in the most trite manner by frequencies, so it is no great miracle that stable frequencies should emerge as the conclusion. Nonetheless it should be stressed that, despite many comments by Popper himself that suggest just this, the propensities we are concerned with in such activities as games of chance are not fundamentally propensities to yield frequencies but propensities to produce single events. There do indeed exist propensities to produce frequencies, but these are explained in terms of the more fundamental propensities.

The propensity interpretation of probability is inescapably metaphysical, not only because many propensities are postulated that are not open to empirical

the Philosophy of Science, **46** (1995), pp. 1–32, is not a criticism of the propensity interpretation.

evaluation but, more importantly, because the idea of necessity, or of law, or of compulsion is built into the theory from the start in a way that, as we saw in section 1, it need not be built into *prima facie* deterministic theories. The modal element in the propensity interpretation cannot be dismissed as idle metaphysics, for without it there is no objective interpretation of single-case probability at all. Here, as elsewhere, it is rash to suppose that non-trivial probabilities are as well behaved and uncomplicated as trivial probabilities are.[23]

Metaphysical indeterminism accordingly explains statistical stability by pointing to stable propensities. How does metaphysical determinism fare? The express purpose of the argument known as Landé's blade, which is discussed in sections 29f. of *OU*, is to show that determinism can give no explanation of statistical stability other than a completely trivial one; statistical stability is simply postulated. There is, that is to say, no deterministic explanation of stability. This argument is really the only argument against metaphysical determinism itself to appear in *OU*, and Popper set some store by it. Later Watkins echoed this emphasis on the argument, and expounded his understanding of it at some length.[24] But some ten years later, in response to unsympathetic criticism from the late J. L. Mackie, he retracted it in

[23] On this point see Miller, *Critical Rationalism*, pp. 104f. For further discussion and defence of the propensity interpretation, see ibid., Chapter 9.6.

[24] See section 2.4 of J. W. N. Watkins, 'The Unity of Popper's Thought', in P. A. Schilpp (ed.), *The Philosophy of Karl Popper, The Library of*

the form originally given, declaring it invalid, and going so far as to suggest that since the argument 'had not ... attracted much notice outside the Popper circle, it might seem that a quiet burial, perhaps in the form of a brief note of retraction, is all that is called for'.[25] But Watkins did not content himself with quiet interment. In addition to stating why he thought the argument to be invalid, he proposed a variant of it that he considered to be valid, and indeed to be 'perhaps the strongest argument, outside quantum mechanics, against physical determinism'.[26] Although I do not have space to consider the matter in detail, I should like to say something on this topic. I too think that the argument is invalid, and that Landé's blade no longer cuts much ice, but not quite for the reason that Watkins diagnoses. If I am right, then Watkins's variant of the argument is also invalid.

Landé's blade is an imaginary device consisting of a vertical chute down which billiard balls can be delivered on to the edge of a blade. Although both Popper (*OU*, section 29) and Watkins, following Landé, pay some attention to the positioning of the blade so that the distribution of the deflected balls on the two sides is 50:50, this does not, in my opinion, have any bearing on the argument. All that matters is that the blade is not so skew that all the balls fall to the same

Living Philosophers, volume XIV (La Salle: Open Court, 1974), pp. 371–412.

[25] See p. 13 of John Watkins, 'Second Thoughts on Landé's Blade', *Journal of Indian Council of Philosophical Research* **II** (1985), pp. 13–19.

[26] Ibid.

side (or, indeed, miss it altogether). In these circumstances, experience tells us, we shall find that the distribution of balls to left and right, whatever distribution it is, is random and seemingly unpredictable, but statistically stable. How is this stability (not the precise distribution, which may depend on many special factors) to be explained? Metaphysical determinism, according to Landé, has only one route to an explanation, and that is to postulate hidden asymmetries at each impact of a billiard ball with the blade. (This is indeed what we quite correctly do when asked to explain the statistical stability of throws with a die or with a coin.) Some of these asymmetries—small draughts, spots of grease, a bit of English —are left-inclining, some right-inclining, and they are themselves distributed in what we describe as a random manner. Each is causally sufficient (we may suppose) to determine which side of the blade the ball will fall. Now Landé simply accepts this answer, and proceeds to ask for an explanation for the statistical stability of the left-inclining and the right-inclining causes. Determinism seems to be in no better position than before, and to be compelled to postulate another earlier sequence of asymmetries that is also statistically stable, and so on. The argument ends only when it is conceded that appropriate asymmetries must have been introduced into the universe at the very beginning, magically attuned to yield statistical stability aeons later. Alternatively, determinism offers no explanation but a concomitance of accidental occurrences. Whichever line the determinist retreats to, the indeterminist seems to have much the better of the encounter. The explanation of stability in terms of propensities may not be profound, but it is not empty, and it is not mystical.

The trouble with this argument is that its conclusion is false. The theory of dynamical chaos provides us with hosts of examples of random (or, perhaps better, pseudorandom) sequences generated deterministically. It is important to appreciate that there is no pseudorandomness in the sequence of values of the logistic function f mentioned in the previous section; this sequence would very quickly fail a test for randomness, since (for example), if $f(t)$ =0.10 then $f(t + 1) = 0.36$. But if we consider only whether the value of $f(t)$ is greater or less than 0.5, we shall obtain a sequence that has every appearance of being random. Dynamical chaos appears to provide just the kind of deterministic explanation of statistical stability that the argument of Landé's blade declares to be impossible.

Why is the argument of Landé's blade invalid? To see this, let us take an excursion into science fiction and graft the logistic function directly on to the blade, supposing that $f(t)$ measures some time-dependent characteristic of the blade. Suppose too that when a ball strikes the blade at time t, the value of f changes (by some mechanism that need not concern us to $4f(t)(1-f(t))$). If at the time t of an impact $f(t)$ >0.5, then the ball is deflected to the right, while if $f(t)$ <0.5, the ball is deflected to the left. (For simplicity, suppose that $f(t) = 0.5$ is impossible.) It is apparent that this purely deterministic piece of apparatus will generate a pseudorandom sequence of outcomes. There are, as the determinist says, small asymmetries at the edge of the blade that are responsible for each deflection. But it does not follow that each asymmetry is the causal outcome of a significantly earlier asymmetry, and that determinism must postulate sequences of antecedent asymmetries in the distant past. The causal chains do not stretch vertically back in time parallel to the chute, as it were, but horizontally along the sequence of previous changes of value of f. The mistaken assumption in Landé's argument is the assumption that because the falls to one side or another of the blade are statistically independent, therefore they and their antecedent causes are also statistically independent. This is simply false. Determinism can explain a statistically independent sequence of outcomes as the rigidly deterministic consequence of a statistically dependent sequence of antecedent conditions.

In his discussion of Watkins's original treatment of Landé's blade, Mackie suggests that the error lies in confusing randomness (that is, statistical stability) in the initial conditions, and randomness (that is, indeterminism) in the laws of development, and that the determinist can easily live with the former, though not of course with the latter.[27] Watkins seems largely to agree.[28] But that is to miss the gist

---

[27] See pp. 369f. of J. L. Mackie, 'Failures in Criticism: Popper and his Commentators' [Review article of Schilpp, *The Philosophy of Karl Popper*, *The British Journal for the Philosophy of Science*, **29** (1978), pp. 363–375.]

[28] He seems even to hold that the indeterminist too has to accept that 'the set-up is beset by various little asymmetries and disturbing influences, but these do not have any systematic bias towards either left or right' ('Second Thoughts on Landé's Blade', p. 16). This is surely a mistake. Indeterminism assumes at most the constancy of the propensity of balls to fall one way or another, and is unbothered by whether or not there are asymmetries present.

of Landé's blade, which asks for an explanation of statistical stability in nature. Watkins goes on to assert that '[a] confluence of rigidly deterministic causal chains that are largely independent of one another may yield a chaotic collective result; but this does not mean that a determinist has to postulate a chaotic ancestral state of which the present chaos is the descendant';[29] and in support of this assertion he gestures in the direction of gas dynamics to provide a counterexample. But he simply assumes at this point that gas dynamics (actually billiard ball dynamics, but it comes to the same thing) is deterministic. Since it was the purpose of Landé's argument to show that, because independent causal chains emanating from a starting point showing no statistical stability will not deterministically generate statistical stability, gas dynamics must be indeterministic, Watkins fails, this second time round, to come to grips with the central issue.

The reason that the argument of Landé's blade fails is that the determinist need not accept that independent outcomes are the progeny of independent acts of generation. But it hardly suffices for the determinist simply to note this and pass on. In cases where statistically stable outcomes are apparently produced from regular and non-random beginnings, we need details of what the non-linear process responsible is. My grafting of the logistic function on to the blade was, as I admitted, no more than science fiction. If the determinist cannot do better than that, then he has

[29] Ibid.

given us no more than a promissory note of determinism, and is maintaining an explanatory theory in a form truly beyond challenge. To be sure, in the case of a macroscopic chute and blade, as in the tossing of a die or a coin on to a soft surface, I accept, as everyone else does, that there are usually small asymmetries present in the apparatus (and, indeed, that these can with care be eliminated). But I do not think that this is obviously so in the case, say, of incidence of photons at a half-silvered mirror.

The upshot of this is that I remain sceptical of Watkins's attempt to provide a valid analogue of Landé's blade, in the case in which the sequence of outcomes is not a statistically stable string but an intentionally meaningful one such as a conversation. Why, Watkins asks the determinist, should the antecedent conditions 'have been so nicely geared to ... [the speaker's] needs one billion years later?'.[30] The answer that the determinist will offer to this is that there is no relevant sequence of antecedent conditions one billion years before the conversation. The relevant initial conditions are all there in the conversation itself. There is no sequence of independent labial and lingual movements in need of explanation by antecedent causes, but a plainly dependent sequence that could in principle be the outcome of some deterministic evolution from a single starting point. I have no disagreement with Watkins concerning the unsatisfactoriness of such a glib answer. Unfortunately I am unable to see how considerations similar to those of Landé's blade are able to denounce it.

[30] Ibid., p. 18.

4. A World of Propensities

Lying behind most discussions of determinism is an interest in human freedom and creativity. As his discussion in 'OCC', sections VII–IX, makes clear, Popper is one philosopher who takes the deterministic threat very seriously, and is under no illusions that there would be any authentic freedom in a physical world that was fully determined. More, he has emphasized how important it is for us that the world should not only not be determined at the level of physics—in the domain that, from the early 1970s onwards, he referred to as world 1—but that it should be causally open to other influences; especially those from world 2—the world of mental activity—and (through the intercession of world 2) those from world 3—the world of abstract human creations, especially problems and theories ('OCC', section X and 'IINE'). Our theories, created by us and encoded by our efforts in structures in world 1, are thereafter sustained there without direct intervention from us. On being rethought, they are able to bring about effects in world 1 itself, for example the construction of new machinery to the designs of an abstract blueprint. But even this turns out not to be enough to make sense of human cultural achievement (in the broad sense of that term). We are not sophisticated machines responding to stimuli but, like all living things, problem-solvers attempting to make our way—and I really mean make our way—in the world. The existence of causal indeterminacies is essential to these attempts of ours, but a mere menu of abstract possibilities from which we may be served is not sufficient. At the very least the available possibilities must be endowed with some potential of realization, some activity. It is here, I think, that Popper saw some metaphysical rewards to be gained from the postulation of active propensities, which may be harnessed by us in order to drag ourselves forwards into the open future. For propensities give promise of providing the much needed 'medium betwixt chance and an absolute necessity'.[31]

From the beginning of the enterprise of interpreting most physical probabilities as propensities of the world (or in special cases, parts of the world) to develop in certain ways Popper had likened these propensities to Newtonian forces. To be sure, they combine in very different ways from Newtonian forces, and they can be annulled in a way that Newtonian forces cannot be. The greater force does not always prevail.[32] But they are supposed to have the same kind of potentiality as these forces. In *AWP*, published in 1990, he tells us (*AWP*, p. 9) of the propensity interpretation that

> it was only in the last year that I realized its cosmological significance. I mean the fact that we live in *a world of propensities*, and that this fact makes our world both more interesting and more homely than the world as seen by earlier states of the sciences.

---

[31] David Hume, *A Treatise of Human Nature* (1739), edition of L. A. Selby-Bigge (1888), p. 171, quoted in 'OCC' (*OK*, p. 227).

[32] This has been a standard criticism of the propensity interpretation of probability. See, for example, D. H. Mellor, *The Matter of Chance* (Cambridge University Press, 1971), p. 158, and Anthony O'Hear, *Karl Popper* (London: Routledge, 1980), pp. 136f. The reader is referred to the text to note 23 above.

I think that the phrase 'in the last year' does less than justice to some of the speculations in 'A Metaphysical Epilogue', Chapter IV of *QTSP*, but let that pass. In any event, what we have here is an attempt to see the cosmos, including ourselves, as the result of the realization of propensities and of the emergence of new possibilities. This question of the emergence of novelty was dominant in Popper's thinking about indeterminism, to the extent that Bartley, the editor of *The Postscript*, was moved to promote the book under the general banner that something can emerge out of nothing, challenging the traditional wisdom that there is nothing new under the sun.[33] We may take the following two successive paragraphs from pp. 18f. of *AWP* as exemplary:

> This view of propensities allows us to see in a new light the processes that constitute our world: the world process. The world is no longer a causal machine—it can now be seen as a world of propensities, as an unfolding process of realizing possibilities and of unfolding new possibilities.
>
> This is very clear in the physical world where new elements, new atomic nuclei, are produced under extreme physical conditions of temperature and pressure: elements that survive only if they are not too unstable. And with the new nuclei, with the new elements, new possibilities are created, possibilities that previously simply did not exist. In the end, we ourselves become possible.

[33] See the Editor's Foreword to *QTSP*, p.xiii.

I should like to end my discussion of Popper's views on propensities and indeterminism with a very brief look at the problem of how new possibilities can emerge, and in particular whether such emergence of novelty is compatible with the vision of the world as an all-encompassing field of probabilistic propensities. For at first sight it does not seem that there can ever be genuinely new possibilities.

There is one uncontroversial sense in which new possibilities can come into existence, even in a deterministic world. For what is possible at one time may not have been possible at an earlier time, and hence the passage of time itself may suffice to make possible what previously was not possible. On the day I was born it was impossible that I should give a lecture at the Royal Institute of Philosophy. But this occurrence; once impossible, has now all but conclusively been shown to be possible. To take care of this, a determinist would insist that the occurrence claimed once to be impossible, later possible, must be provided with a date, though perhaps not a very precise one; and that once this is done it is clear that there always existed a possibility (at least since I was born) that I should give a lecture at the Royal Institute of Philosophy in November 1994. Those who talk about the emergence of new possibilities mean something more than that. Moreover, they mean something less than the emergence of new logical possibilities. No doubt there are some who would want to say that logic is only something human, and there is no reason why it should not expand or contract. But Popper was not among them, and I have no intention of pursuing the question in that direction.

What is surely meant in *AWP* by saying that new possibilities can be created is not merely that new possibilities can be created, but that new propensities, new forces—or centres of force—can be created. The problem with this picture is that it seems to clash with the identification of propensities and probabilities. If a dated event has no propensity, and hence a zero probability of occurring, then that probability can be raised to a positive probability only by the occurrence of another event of no probability, or no propensity. This very familiar consequence of probability theory, that only events of zero probability can alter the probabilities of other events of zero probability,[34] is one that Popper himself wielded on a number of occasions against others (see, for example, *LSD*, appendix *vii, especially p. 364). It is a little surprising to see his own ideas apparently at the mercy of the same result. For what it means is that new non-zero propensities cannot emerge without the earlier actualization of events of zero propensity; and this looks very much like saying that new possibilities can come about only if something impossible happens first.

Popper certainly intends zero probability to indicate the absence of propensity. He writes, for example, that 'zero propensities are, simply, no propensities at all, just as the number zero means "no number" . . . a propensity zero means *no* propensity' (*AWP*, p. 13). But despite the close connection between propensity and possibility, there is no

---

[34] If $p(a, c) = 0$ then $p(ab, c) = 0$ by B1 and (18) of *LSD*, appendix *v. By M2, $p(a, bc)p(b, c) = 0$. Thus b can raise above 0 the probability of a, given c, only if b's own probability, given c, is 0.

suggestion that zero probability implies impossibility. On the contrary, he writes elsewhere that 'zero probability... means, in the case of random events, *a probability which may be neglected as if it were an impossibility*' (*RAS*, p. 380). Herein, I think, lies the solution to our problem.

We must recognise not only that zero propensity does not imply impossibility, but that there exist all the time possibilities whose propensity to occur is strictly zero. Popper himself nudges us in this direction with the remark that propensities are '*more than mere possibilities*' (p. 12). Such events have no propensity to occur, but they may occur nonetheless—by accident, as it were. Indeed, accidental or chance occurrences seem to me to be exactly what we need here. To misuse an example of Aristotle's, a tile slides off a roof and strikes a passer-by. Two unrelated causal chains, as we put it, coalesce by chance. There was no propensity at all for such an occurrence, but it was not abstractly impossible. It is the kind of occurrence that immediately brings into being new propensities, propensities that were previously zero. Serendipity is more significant than is sometimes thought. Like Popper, I find incredible the idea that at the beginning of the universe, or after the first three minutes of it, there was any propensity for the *St John Passion* eventually to be written, or for *The Night Watch* to be painted, or for the Parthenon to be built, even though these achievements were abstractly possible. If we admit that some things may happen without there having been any propensity for them to happen, we can happily accept that for most of recorded and unrecorded history there were simply no such propensities.

The world is therefore not a world run wholly by the operation of propensities. It contains many chance events too, events for which there never was any propensity of occurrence. It is these events that somehow we have learnt to take advantage of. I need hardly insist that there has to be much more to the story than this.

7   Popper on Determinism

PETER CLARK

1.   Introduction

There is no doubt at all that the issue of determinism versus indeterminism was a central, dominating theme of Popper's thought. By his own account he saw his criticism of the thesis of determinism as crucial to his defence not only of the reality of human freedom, moral responsibility and creativity but also as equally fundamental to his account of human rationality and to his theory of the content and growth of science as an objective, rational and most importantly *demonstrably* rational enterprise. Consequently a great deal of his writings discussing both the content and methodology of the natural and the social sciences alternately bear upon and presuppose his defence of indeterminism.

Like many distinguished philosophers before him, he held that notions crucial to the way we ordinarily see ourselves as rational agents would be rendered completely otiose if the thesis of physical determinism were globally true. The truth of determinism he urged would entail that we inhabited a nightmare world. Commenting approvingly on the opening passage of Arthur Holly Compton's *The Freedom of Man* Popper wrote as follows:[1]

> Compton described here what I shall call 'the nightmare of the physical determinist'. A deterministic physical clockwork mechanism is, above all, completely self contained: in the perfect deterministic physical world there is simply no room for any outside intervention. Everything that happens in such a world is physically predetermined, including all our movements and therefore all our actions. Thus all our thoughts, feelings, and efforts can have no practical influence upon what happens in the physical world: they are, if not mere illusions, at best superfluous by-products ('epiphenomena') of physical events. ('OCC', in *OK*, p. 217)

Later he returns to the theme of the nightmare of determinism. He says (*OK*, p. 222):

> It is a nightmare because it asserts that the whole world with everything in it is a huge automaton, and that we are

---

[1] Popper remarks of the problem raised by Compton there that it is the only form of the problem of determinism worth discussing seriously. He characterizes it as 'the problem which arises from a physical theory which describes the world as a *physically complete* or a *physically closed* system'. He goes on, 'By a physically closed system I mean a set or system of physical entities, such as atoms or elementary particles or physical forces or fields of forces, which interact with each other—and only with each other—in accordance with definite laws of interaction that do not leave any room for interaction with, or interference by, anything outside that closed set of physical entities. It is this closure of the system that creates the deterministic nightmare.' (*OK*, p. 219. The footnotes to this passage are omitted; the italics are Popper's).

> nothing but little cog-wheels, or at best sub-automata, within it.
>
> It thus destroys, in particular, the idea of creativity. It reduces to a complete illusion the idea that in preparing [The Compton Memorial] lecture I have used my brain to create something new.

However, again like many philosophers, he was fully aware that no matter how good the arguments might be in favour of indeterminism, mere indeterminism was not enough to secure a foundation for human rationality, freedom and creativity thereby escaping from the nightmare of determinism. Indeterminism would provide at most merely a necessary condition for the applicability of those concepts. He devoted much effort to trying to provide at least a sketch of a theory as to what extra conditions were needed over and above the satisfaction of the condition of indeterminism to provide an adequate account of freedom and rationality. Here his two main contributions are independently well-known, they are the idea of 'plastic control' and his theory of what he called World One (the world of physical objects and processes) and World Three (the world of abstract objects and theories) interaction via World Two (the psychological world of beliefs and desires).[2]

---

[2] 'Epistemology Without a Knowing Subject', *OK*, pp. 107–152 and 'On the Theory of the Objective Mind', *OK*, pp. 153–190. Popper undoubtedly accepted the reality of abstract objects, like numbers and sets for example but his account of them and of how we come to know them was certainly neither Fregean nor constructivist. World Three contains theories which are presumably to be thought of as free creations of the human intellect. But it also contains abstract objects like

The seriousness and depth of Popper's concerns with the issue of determinism are surely not to be doubted. Nor is it to be doubted that he made a number of important conceptual innovations to the study of the problem, particularly in connection with the explanation of statistical stability and the theory of probability and the interpretation of that theory as it is used in statistical mechanics and quantum mechanics. Indeed I shall argue that Popper's most important contribution in this area is his introduction of the idea of 'propensity' as an explanation of statistical stability and it is to his credit that he saw with real clarity the extraordinarily deep metaphysical problem posed by the fact that the world exhibits stable, non-trivial statistical regularities. The idea of propensity as a property of experimental arrangements, though not without very considerable interpretive difficulties is a very important conjecture designed to solve a deep problem. Further it has nothing whatever to do with Popper's well-known criticisms of very strong claims about the predictability of events, claims which have, however, little if anything to do with determinism.

What however can be doubted is whether Popper's criticisms of determinism succeed and indeed whether a whole class of arguments deployed against that thesis could have been thought relevant to the avowed concerns which according to his own account motivated them. There are essentially two central difficulties with Popper's account of determinism. The first is his dismissal as he puts it as a 'mere

number which are not as Frege thought to be conceived of as logical objects nor are they to be conceived of as entirely free creations.

verbal quibble' of the Principle of Alternative Possibilities and the second is his systematic stress on the analysis of so-called 'scientific' determinism as opposed to metaphysical or ontological determinism.

2. Alternative Possibilities and Predictability

In the Introduction, written in 1982, to one of his major sustained works on the issue of determinism (*OU*, pp. xix–xxii), Popper introduces with admirable clarity what he sees as the main motivating problem. Common sense and a common sense reading of science seems to tell us (1) that every event is caused by some preceding event and that every event (and therefore the events involved in performing actions) could be anticipated on the basis of sufficient knowledge of all preceding events. On the other hand, common sense also tells us (2) that sane persons can freely choose between alternative possibilities for action and that therein lies their responsibility. Now Popper denies the compatibilist claim that the inconsistency between (1) and (2) is merely apparent. He writes (*OU*, p. xx) 'The arguments on which this [the compatibilist] position is based are, however, largely verbal. They depend upon the verbal analysis of the meaning of such words as "free", "will", and "action"; and upon the analysis of such questions as, "Could I have done otherwise than I did?" these verbal analyses are quite futile and have led modern philosophy into a morass. But there is another approach.'

This is however an extraordinary claim to make, for the Principle of Alternative Possibilities, the claim that one is free or morally responsible for what one has done only if one could have done otherwise than one did in exactly the same circumstances, is a key principle which connects the thesis of determinism with the issue of freedom. This is because if determinism is true the condition for freedom and responsibility appears never to be satisfied. If determinism is true the agent could not have acted differently in exactly the same circumstances and so could not have done otherwise. Popper might be right to dismiss compatibilism, but that cannot be because the Principle of Alternative Possibilities is a 'verbal quibble', which it manifestly is not.

He often argues that the main issue of determinism is an issue of the completeness or closure of the system under physical description (*OK*, p. 219), but this is really just another way of putting the same point. If the physical description of some system is causally complete and deterministic then any other information must be redundant or contradictory, so if the physical conditions are exactly the same, then the subsequent events and so actions must be the same. Either way of looking at it yields the conclusion that it is the Principle of Alternative Possibilities which renders the connection between the content of physical theory and the issue of human freedom.

It also makes clear that predictability of human action is not the main issue. One's decisions might be predictable, but nevertheless regarded as free and responsible, similarly one might consider the output (the decisions) of some automata to be determined, perhaps for some very high level theoretical reason, but nevertheless it be the case that actual outputs were highly unpredictable because of some extreme dependency upon initial conditions whose exact values were even in principle unknowable by us.

Now what was the other approach advocated by Popper? It was to follow not Hume but Laplace. Certainly it is true that one of the very refreshing and innovative aspects of Popper's discussion of determinism following that of Laplace is the concern he shows for tying the discussion down to what science actually tells us about the world, but when he actually does this he sometimes departs so far from the actual content of the physical theories he is discussing, by *adding* to them very strong claims about predictability, that the idea of examining what the theory really says completely disappears in arguments which manifestly transcend the content of the theory in question.

This certainly does not prevent him from making some deep observations on what looks like support for the deterministic thesis from some scientific theories, and he does show that it turns out not to be quite so clearly supportive as appearances might lead one to suspect in at least one specific case, but his constant returning to the notion of predictability as distinct from determinism does undermine almost entirely any claim as to the relevance his negative arguments have to the central motivating concern, that of human freedom and responsibility. Indeed some of the claims he makes based on the untenability of very strong claims about the predictability of all events including future events in the history of the predictor raise new problems about human rationality, indeed of the rationality of science itself.[3]

---

[3] These arguments have been effectively criticized by Peter Urbach. See his 'Is Any of Popper's Arguments Against Historicism Valid?', *British Journal for the Philosophy of Science*, **29** (1978), pp. 117–130.

The arguments of the *Poverty of Historicism* to the effect that social trends and forces are unpredictable and that the growth of knowledge is inherently unanticipatable seem to fly in the face of ordinary human experience and apparently throw into doubt the rationality of activities which form the basis of social structures which are fundamental to modern society. Modern developed societies do behave in predictable fashion and we have good reasons for believing that to be so, otherwise what would be the rationality of spending huge quantities of resources in time and money in trying as every serious firm or enterprise (from insurance companies to government departments) does to anticipate the environment in which it will be competing in the future. Further, what would be the rationality of the activities of the major research councils in science (not just in technology), when they appraised research grant applications and rated them worthy of support or otherwise if it really were impossible in principle to anticipate how knowledge will grow. The rationality of the activities of the science research councils must, in part, hinge upon the correctness of the conception that it is possible to predict to some extent at least which projects are likely to be successful, and that fundamentally involves predicting how knowledge will evolve. What is remarkable is not argument which purports to show how in some theoretical limit of complete exactness prediction of the growth of knowledge is impossible but how and why the work of research councils, which is so fundamental to scientific activity, can achieve what it does achieve.

243

3.   The Analysis of Determinism

The constraints that Laplace and Popper following him actually impose are very strong. Essentially in modern terms Laplace implicitly imposes three conditions: first that an *analytic* solution to the equations of motion for a system obeying Newtonian mechanics *expressible as a closed formula exists* for *all* initial conditions (i.e. as Laplace has it his demon would provide solutions which 'would embrace in the same formula the movements of the greatest bodies of the Universe and those of the lightest atom'). The second constraint amounts to the requirement that every solution to the equations of motion for such a system shall be *effectively computable* in the data, that is, given as initial conditions the position and momentum (the values of the state variables) of all the particles in the Universe at some given instant, the values of the state variables at any subsequent instant shall be effectively computable functions of the state variables at the arbitrarily chosen initial instant (i.e. as Laplace has it that the solution should enable one to 'foresee those [phenomena] which given circumstances ought to produce'). The third constraint is one of the complete accessibility of the data of initial conditions, that is, it is to be assumed in the formulation of the deterministic claim that it is always possible (classically at least) for experimental evidence to fix an exact real number as the value of each state variable at an instant.[4] No doubt if these very strong conditions were to

---

[4] The quotations from Laplace occur in the famous passage from his *A Philosophical Essay on Probabilities* (1820), p. 4; Dover reprint, 1951.

obtain then Laplacean global predictability would obtain, but there is no reason at all to tie the claim of determinism to a thesis of global predictability.

It is clear that this latter could fail for all sorts of reasons, for example, the failure of any functional dependencies to be effectively computable (Who is to say *a priori* that every law of nature must involve only functions recursive in the data?).[5] It may be that unlike the capacities of Laplace's demon, some such dependencies are simply beyond our, or the very best computers' capacities to survey or compute. Indeed Popper was able to show in his 1950 paper (*IQP*; and also *OU*, pp. 64–77) that for purely logical reasons no computer would be able to predict *all* the future states of a universe of which it itself was a part. This does not of course mean that Laplacean determinism is contradictory, for a simple model of it is provided by a universe consisting of a single particle moving inertially against a background of Newtonian absolute space. A more interesting model is provided, as Earman has noted,[6] by the only known general solution to the N-body

---

[5] Indeed there are some unequivocal examples of the failure of recursiveness in classical physics, see particularly Marian Pour-El and Ian Richards, 'Non-Computability in Analysis and Physics', *Advances in Mathematics*, **48** (1983), pp. 44–74.

[6] Earman's excellent *A Primer on Determinism* (Dordrecht: D. Reidel, 1986) is by far the best treatment of determinism and its relation to physical theory in the literature. Pollard's example mentioned above is noted by Earman, Ibid, p. 54. It is given as an exercise in Pollard's *Mathematical Introduction to Celestial Mechanics* (Carus Mathematical Monographs, Vol. 18, 1976), p. 59, Exercise 1.3.

problem, a universe consisting of N material particles moving under an attractive law of force acting along the line of centres but proportional to the distance of separation.

However Popper is surely right that one cannot say categorically that classical mechanics is deterministic (*OU*, pp. 43–44).[7] It is in his phrase only *prima facie* deterministic, and this is true even when, a very much weaker characterization of determinism is given than is used by Popper.[8]

---

[7] It is important to recognize that long before chaos theory was fashionable in philosophical circles Popper drew attention to a very important property of some deterministic systems, namely, how their large scale macroscopic states were so finely dependent on initial conditions. It is an open question of considerable difficulty as to what extent, if any, this instability can be used to mimic indeterministic, or even much more strongly, random behaviour. Popper's ideas on the subject, based on classic papers by Hadamard and Hopf, are summarized in *OU*, pp. 39–40 and *QTSP*, pp. 104–116.

[8] Popper is surely right in his contention (*OK*, p. 220) that it is not much use to appeal to well-worn maxims such as 'Every event has a cause' and 'Like causes produce like effects', when trying to formulate the intuitive notion of determinism more formally. This is so first and foremost because the notion of cause plays no role in either the formulation or the content of physical theory, and secondly because the notion of cause is so notoriously vague and context-dependent that it is quite unsuitable to employ it to capture the general notion of determinism. Neither would it be sensible to characterize deterministic theories as those which make no reference (or merely trivial reference) to probability, since a physical theory may well involve a parameter defined using a probability measure which itself evolves deterministically with time. The problem with formulating the thesis of determinism is the obvious difficulty of trying to steer a course between triviality on the one hand and far too strong a claim on the other. Thus Russell and Hempel have with great clarity established the charge of triviality against some

This weaker characterization of determinism, is what physicists really mean by determinism. It is that the state at one time fixes the state at future times. What this means for mechanics is the well known fact that because of the existence and uniqueness of solution of the differential equations of motion, the time evolution of a dynamical system (say a system of moving molecules, the familiar billiard balls of a dilute gas) is represented by a unique trajectory in phase space (the space in which a point corresponds to the instantaneous state of the system).

Through each point of the phase space there passes *at most* one trajectory, as a consequence of uniqueness of solution. In any time interval Δt, the point on the trajectory corresponding to the state of the system at t is transformed into the *unique* point on the trajectory corresponding to the state at $t + \Delta t$. But of course during the same time interval *every* other point in the phase space is transformed uniquely by the solution to the differential equations of motion into another point in the phase space. Hence the solutions to the equations of motion define a map from the phase space of the system into itself with the property that as the dynamical system evolves through time, that evolution induces a '*flow*'

syntactic formulations of the condition employing merely the existence of functional dependencies between the time and evolving states of a unique Universe (Russell 'On the Notion of Cause', *Mysticism and Logic* (London: Allen & Unwin, 1918), pp. 132–151.), and involving conditions sufficient for an event to occur (Hempel 'Some Reflections on the "The Case for Determinism"', in S. Hook (ed.), *Determinism and Freedom*, (London: Macmillan, 1958), pp. 170–175.)

of the points of phase space. The flow has the very important property that it characterizes a one parameter (the time) semi-group, which means that if we consider any point in phase space and look at where the flow takes the point in time t and then subsequently in time interval δ, the final position of the point is exactly that to which the flow takes the point in time interval $t + \delta$. This is the familiar *natural* motion of the phase space. (Since the equations of motion of classical mechanics are time-symmetric, the natural motion of the phase space has the structure of a one parameter group.)

As was noted by Montague in his classic paper of 1974 on deterministic theories, this idea can be generalized in a natural way.[9] If we consider an *isolated* (this restriction is essential) physical system and think of the history of that system as the 'graph' of the values of its state variables in phase space (i.e. the n-dimensional space in which the instantaneous state of the system is a 'point') through time, then the system is *deterministic* if and only if there is one and only one possible path consistent with the values of its state variables at any arbitrary time. There are however two basic problems in formulating this intuitive condition

---

[9] Richard Montague, 'Deterministic Theories', in R. H. Thomason (ed.), *Formal Philosophy* (New Haven: Yale University Press, 1974). See also B. van Fraassen, 'A Formal Approach to the Philosophy of Science', in R. G. Colodny (ed.), *Paradigms and Paradoxes*, (Pittsburgh: University of Pittsburgh Press, 1972), pp. 306–366. Montague's definition is not itself without some formal difficulties, see especially G. Hellman, 'Randomness and Reality', *PSA*, **2** (1978), pp. 79–97.

precisely. One concerns the appeal to physical *systems*, and the other, the notion of 'possibility'.

The first difficulty stems from the fact that systems can exhibit deterministic and indeterministic aspects simultaneously. An obvious example is a system obeying classical mechanics if we consider only the temporal evolution of its observational states. Since an observational state will in general have a dynamical image (i.e. the set of phase points compatible with the observational state) containing more than one phase point, two systems obeying the same dynamical laws in the same observational state at any given time may well be found in different observational states at later times. It is precisely this characteristic which gives the deterministic thesis its *hidden variable* aspect; for a claim that a system is indeterministic may rest merely upon our not having obtained a description of the underlying state variables, that is, the theory of the system may be radically *incomplete*.

The second difficulty concerns the introduction of the modality, possibility, and the attendant panoply of real but not actual possible words. One way of avoiding this is to think of the class of all models of the theory. To cut a quite long story short, *a theory T* is said to be *deterministic in the state variables* (say $\delta_1, \ldots, \delta_n$) if any two standard models, or histories, as they are called, of the theory (i.e. any two relational structures which are models of the theory which agree at some given time, agree at all other times). In short the constraint entails that for a deterministic theory if two histories (or models) have identical states at one time then they have identical states at all times. A physical system may be said to be deterministic in the state variables $(\delta_1, \ldots, \delta_n)$ when its history realizes (or satisfies) a *theory* deterministic in the state variables $(\delta_1, \ldots, \delta_n)$. This characterization of a deterministic theory and deterministic system fits very well with classical, relativistic and quantum mechanics. It has the further advantage of not relying on the essentially extraneous notion of predictability at all.

Whereas Popper was right in suggesting that Newtonian mechanics was only *prima facie* deterministic, his analysis of the problem with respect to special relativity was again marred by extraneous considerations concerning predictability and what he called the 'block universe' model of Minkowski space—time. Briefly put, it turns out that special relativity is more deterministic than classical mechanics because the light postulate rules out the counterexamples to determinism that can arise in the classical case.[10] But perhaps what is more interesting is not in this context the details of the case, but what Popper could have thought was established by the argument he gives. He remarks (*OU*, p. 59) concerning the verdict on determinism of special relativity that: 'But as a consequence of this, the future becomes "*open*" to us in the sense that it cannot be fully predicted by us, while the past is closed; that is to say, the asymmetry is of the kind which I have tried to establish.' But what sense of openness, if any, would this show?

If the sense of openness is to be associated with newness, genuine possibility, creativity, past states failing

---

[10] Earman, *Primer on Determinism*, chapter IV.

to fix future states, or the reality of time, what relevance is it that the future cannot be predicted? For it might very well be the case (indeed is the case in the special theory providing the time slice forms a Cauchy surface) that while possibly unpredictable it is causally fixed by antecedent events. A possibility not considered or anticipated by me is not thereby a *new* possibility. Again it is far from clear how Popper's argument relates to his fundamental concerns.

A similar difficulty appears with respect to Popper's criticism apparently directed at Einstein of the four dimensional block universe interpretation of Minkowski space–time of special relativity. It was one his criticisms of such a view that:

> the future, being causally entailed by the past, could be viewed as contained in the past, just as the chick is contained in its egg. Einstein's determinism made it *completely* contained in the past, in every single detail. The future became therefore *redundant*. It was *superfluous*. (OU, p. 91)

It is certainly a recurrent theme in Popper's thought that determinism entails that the future is 'contained' in the past. No doubt in some sense it does so, but that in no way justifies the additional metaphors of 'redundant' and 'superfluous'. No doubt the responses of my computer are fixed in complete detail by manufacture, history and programming, but that hardly renders those redundant or superfluous. More to the point, however, is the question why a so-called block universe model is supposed to be committed to determinism? Surely this cannot be right. One could have a four dimensional block universe model of an indeterministic Universe all that that would require would be that the events comprising a suitably chosen time slice of that Universe did not fix, together with the appropriate laws of nature for that universe, the events associated with later times. Block universe models may have implications for the reality of time or change, or more exactly, for the reality of tense, but they have no implication as to the truth of determinism.

4.   Propensity

No consideration of Popper's views on determinism can be complete without an examination of his most important contribution to the study of metaphysical determinism, namely the propensity interpretation of the calculus of probability. The connection between the two conceptions is easy to see, for according to Popper 'propensities can be accepted as physical realities (analogous to forces) only when determinism has been given up' (QTSP, p. 105).[11] But this claim raises a very considerable difficulty in an area which Popper wishes to deploy the propensity theory to resolve a conceptual problem, namely, in respect of how to understand physical probability as it occurs in statistical mechanics. Popper says:

> Today I can see why so many determinists, and even
> ex-determinists who believe in the deterministic

[11] An extensive account of the propensity theory can be found in Popper's RAS, Part II.

character of classical physics, seriously believe in a subjectivist interpretation of probability: it is in a way, the *only reasonable possibility* which they can accept; for objective physical probabilities are incompatible with determinism; and if classical physics is deterministic, it must be incompatible with an objective interpretation of classical statistical mechanics. (*QTSP*, p. 105)

Here we have a very serious consistency problem indeed, for it looks as if we cannot accept both the determinism of the underlying part of classical mechanics and the existence of real physical probabilities as ordinarily understood in classical statistical mechanics.

Clearly Popper is inviting the reader to contrapose. The objective interpretation of statistical mechanics is the right one, so classical physics must be indeterministic. The argument then is as follows: since the statistical assertions of statistical mechanics are just as objective as the 'mechanical' assertions of that theory and since Popper argues those assertions cannot be treated as such if the underlying dynamical theory is deterministic, then the underlying dynamical theory must after all be indeterministic despite its *prima facie* deterministic character. Now there is no need in *this* context to become embroiled in difficulties with the notion of objectivity. Simply let the dynamical postulates (Newton's laws) and the statistical postulates be treated on a par, as correct and jointly exhaustive descriptions of the (classical) physical situation, the transition to quantum statistical mechanics does not resolve any of the difficulties discussed here.

Popper uses an example from non-equilibrium statistical mechanics to make his point. In order to explain an irreversible approach to equilibrium, we have to assume that the measure of the set of possible initial states which produce 'pathological' (i.e. non-approach to equilibrium) behaviour is zero, and the measure of the set of possible initial dynamical states which do eventually yield equilibrium is one. Now this probabilistic hypothesis, that the set of 'pathological' states has measure zero, has to be physically interpreted as a claim about *propensities*, but propensities only obtain where exactly the same situation or 'state' may yield, in the time evolution of a system, different subsequent states; but that is incompatible with *determinism*. But the indeterminism of observational states is trivial here. Whatever propensities are present must be grounded in the *dynamics* (that which in complete exactness controls the behaviour of the system). So, to have dynamical propensities of a non-trivial kind (not always zero or one), we must have, so the argument goes, dynamical indeterminism. However, in the context of classical statistical mechanics there is no other place to appeal to but classical mechanics, so giving up determinism must be the same as saying, classical mechanics is after all indeterministic. But in the context discussed this claim is manifestly false.

Paradigmatically, as employed in standard formulations of statistical mechanics, the underlying mechanics is deterministic. This is because the equations of motion for the dynamical system (the canonical Hamiltonian equations, where the Hamiltonian is independent of the time) are such that we have a system of first order differential equations whose solutions *always exist* and are *unique*, given an arbitrary initial point at some instant t. This is precisely as required by Montague's characterization, or indeed any intuitively acceptable one. Since both existence and uniqueness of solution of the differential equations of motion obtain, the dynamical states of the system are determined. Since we might very well wish to agree with Popper that the physical probabilities occurring in classical statistical mechanics are as objective as the dynamical constraints, but disagree completely with him that the underlying mechanics in *this context* is indeterministic, we are forced, accepting the validity of his argument, to deny that 'objective physical probabilities are incompatible with determinism'. This seems an inescapable conclusion *if* one wishes to deploy the propensity theory in defence of objective physical probabilities in statistical mechanics.

Now it might be objected that propensities are introduced here only to solve an initial value problem. That is to explain why in nature we never (or almost never) observe systems which have as their initial condition just one of those special pathological initial states which subsequently evolve away from (instead of towards) equilibrium. In other words the propensity theory is being used to solve the measure zero problem. This is the problem of explaining why measure zero initial states are never (or almost never) observed. But then it is clear that the matter can have nothing to do with interpretations of the probability calculus.

To see that, it is only necessary to do what the propensity tells us and to replace the probabilistic term measure zero by its interpretation, propensity zero, and the problem then becomes, why are propensity zero states not observed? It is trivial and uninformative to reply that it is because they have propensity zero. Rather we must appeal to some fact about experimental arrangements or set-ups which shows that the propensity zero states will not be observed, but then that particular fact which is explanatory will be extra to the propensity theory. Let us grant that it is a peculiar property of experimental arrangements that certain dynamically permitted states do not actually occur, that is have propensity zero: the question remains, why? The propensity theory alone clearly cannot solve the fundamental issue of statistical mechanics, the measure zero problem.

Neither can it solve the fundamental problem of quantum mechanics for as Milne has pointed out,[12] if propensity is to be seen as a property of whole experimental arrangements or set ups, then since each experiment exhibits one and only one of the two aspects of quantum wave particle duality the theory cannot be used to explain the results of the two slit experiment (an interference pattern), as an interference phenomenon of the two distinct types of arrangement where only one of the slits is open and the distribution is that of the arrival of particles passing through the single slits. The two slit experiment is one experimental arrangement, the single slit experiments are distinct incompatible arrangements (no single arrangement can satisfy any two of the three descriptions). Since propensity is characterized by the whole experimental arrangement, one cannot properly regard the results of the two slit experiment as the

---

[12] 'A note on Popper, Propensities and the Two Slit Experiment', *British Journal for the Philosophy of Science*, 36 (1987), pp. 66–70.

interference of the two propensities each associated with one of the single slit experiments.

Neither in application to statistical mechanics nor to the quantum theory does the propensity theory of probability resolve the outstanding conceptual problems of those theories. That of course, in itself, does not mean that as an interpretation of the calculus of probabilities it is flawed.

5.   Hume's Fork

Perhaps the most interesting contribution of Popper to the notion of indeterminism is his answer to Hume's fork, the idea of 'plastic control' (*OK*, pp. 240–250). This idea is difficult and has not been developed extensively in the literature. The difficulty is certainly not one of characterizing what a *partially* deterministic system might be, that is one, not all of the state variables of which, satisfy a deterministic theory. But the idea of plastic control involves much more than partial indeterminism, for Popper suggests that the behaviour of the system as a whole is influenced in its large scale apparently deterministic properties by the behaviour of the indeterministic variables. It is not that the deterministic state variables constrain a range of values within which the indeterministic parameters may range freely, but that the constraints themselves are influenced by the behaviour of those parameters. Given that we are dealing with something distinct from mere partial determinism, which does not present any particular conceptual advance the difficulty is how to make sense of the required sort of 'influence'.

We can put the problem like this: the problem arises because we know that for certain crucial phenomena, like human freedom, mere indeterminism is not enough, so we want influence without determinism, but what sort of influence can that be? It can be only that it partially determines the outcome, which is to say that the outcome is not determined, which is to say that it is indeterministic over some range or in some aspect. But we have already agreed that that is not enough. Thus the way out of Hume's dilemma seems still obscure.

8   Popper and the Quantum Theory

MICHAEL REDHEAD

Popper wrote extensively on the quantum theory. In *Logic der Forschung* (LSD) he devoted a whole chapter to the topic, while the whole of Volume 3 of the *Postscript to the Logic of Scientific Discovery* is devoted to the quantum theory. This volume entitled *Quantum Theory and the Schism in Physics* (*QTSP*) incorporated a famous earlier essay, 'Quantum Mechanics without "the Observer"' (QM). In addition Popper's development of the propensity interpretation of probability was much influenced by his views on the role of probability in quantum theory, and he also wrote an insightful critique of the 1936 paper of Birkhoff and von Neumann on nondistributive quantum logic[1] (BNIQM).

In this paper I will look at some of Popper's arguments in a suitably critical spirit. But he would have applauded this. My great regret is that he cannot respond to this paper with *criticisms* of *my* arguments!

---

[1] In this paper Popper exploited some ambiguity in the Birkhoff and von Neumann paper concerning the distinction between complement and orthocomplement in a lattice, and by making a plausible measure-theoretic assumption concerning their lattice showed that it was actually distributive rather than nondistributive. For a detailed critique of BNIQM reference may be made to E. Scheibe, 'Popper and Quantum Logic', *The British Journal for the Philosophy of Science* **25** (1974), 319–28.

1. Thought Experiments

In 1934 Popper published a short note in *Die Naturwissenschaften* entitled 'Zur Kritik der Ungenauigkeitsrelationen' (KU). This was actually Popper's scientific debut. It contained what Popper later described as 'a gross mistake for which I have been deeply sorry and ashamed ever since' (*QTSP*, p. 15). Appendix *XII to the 1959 translation of LSD reproduces a letter from Einstein written in 1935 criticising Popper's proposed thought experiment. Although he tells us in his autobiography (*UQ*, p. 92) that he defended his experiment against von Weiszäcker and Heisenberg, he immediately concurred with Einstein's criticism. Einstein was Popper's great scientific hero, and the example of the successful novel predictions of general relativity was the paradigm example of Popperian corroboration, although as we shall see later, Popper was in fundamental disagreement with Einstein on the question of determinism in quantum mechanics. Although Popper's understanding of the physics involved in his thought experiment was definitely at fault, as we shall explain shortly, nevertheless there is a definite sense in which Popper had anticipated an important ingredient in the famous Einstein–Podolsky–Rosen (EPR) paper of 1935. One can even speculate that Einstein was influenced by Popper here.[2] Popper himself remarked in a letter to Max Jammer in 1967, with a humility that was more characteristic of Popper in matters of physics than in matters of

---

[2] For historical details see M. Jammer, *The Philosophy of Quantum Mechanics* (New York: Wiley, 1974), p. 178.

philosophy, 'the possibility that a *gross mistake* made by a nobody (like myself) may have had an influence on a man like Einstein never entered my head.'

I shall now briefly describe the thought experiment, but with some simplifications, that makes the evaluation of the experiment, I believe, more transparent.

Popper's idea was to consider the scattering of two particles, call them A and B where A has a sharp momentum and B a sharp position. (In Popper's actual experiment the collision is between a parallel beam and a diverging beam. I am simply proposing using the origin of divergence as the point of collision—what Popper refers to as the Schnittpunkt. This greatly simplifies the geometry of the experiment.)

Referring to Figure 1, A has a sharp incoming momentum $\mathbf{p}_A$ and scatters into a final momentum state $\mathbf{p}_A{}'$. B has initially a sharp position at the point X and recoils into a momentum eigenstate with momentum $\mathbf{p}_B{}'$. An initially unknown momentum, $\mathbf{p}_B$ is absorbed from the B particle, but this can be computed from conservation of momentum, i.e.

$$\mathbf{p}_B = \mathbf{p}_A{}' + \mathbf{p}_B{}' - \mathbf{p}_A \tag{1}$$

Suppose now that we don't know the magnitude of B's final momentum, but only its direction, i.e. we don't know $|\mathbf{P}_B{}'|$. But this can be calculated from conservation of energy, if we have measured $|\mathbf{P}_A{}'|$ as well as the final direction of the scattered A particle. Popper now wants to show how to *predict* by measurement on the scattered A particle not only the recoiling *momentum* of the B particle but also

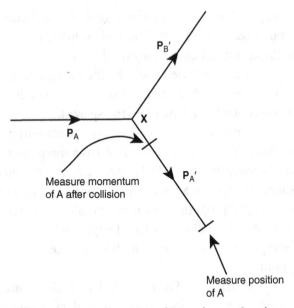

Figure 1. A schematic version of Popper's 1934 thought experiment

its recoiling *position* at any given time. Since the recoil starts at X and we know the momentum and hence velocity of the recoiling particle, as calculated from the measurement of $\mathbf{P_A}'$, we can predict the position of the recoiling particle provided as we know the time at which the collision occurs. To find this, says Popper, all we need to do is measure the *position* of the scattered A particle at a given time *after* the momentum measurement, and retrodict from the known velocity of the A particle after the collision, the time at which it must have emerged from X, i.e. the time of collision.

So putting all this together, by making successive momentum and position measurements on A, we can predict the momentum and position simultaneously for the recoiling B particle.

The flaw in Popper's argument is that measuring the momentum of the recoiling B particle necessarily affects the possibility of retrodicting its position at still earlier times. For example, if we measure the momentum by the Doppler shift of very low frequency radiation, then an accurate measurement of the frequency requires a long wave train, i.e. a long duration of time for the irradiation, during which time the average velocity of the B particle cannot be exactly specified. It is easy to show that this exactly prevents us from retrodicting the position of the B particle *before* the momentum measurement, and hence prevents us from inferring the time at which the recoil particle emerges from X. An alternative proposal put forward by Popper was to measure the momentum by means of a selective filter. But this again would disturb the position of the recoil particle, for example a wave packet of precise location incident on the filter would be indefinitely 'spread out' by the action of the filter in selecting a particular narrow band of wavelengths from the wave packet.

What EPR do, by contrast, is to devise an experiment in which by measuring *either* momentum *or* position on the one particle the corresponding momentum or position can be inferred for the other particle. But since both these measurements cannot be made simultaneously, we cannot *predict simultaneous* values for the momentum and position of the other particle. If we were able to do this a conceptual selection of a 'super-pure' ensemble of the second particle would be possible, with no dispersion in momentum or position, but this would contradict the objective scatter relations, which Popper regards as the correct interpretation of Heisenberg's uncertainty principle. In *LSD*, pp. 254–264 and 301, Popper tries to argue that his thought experiment is not ruled out by violating the objective scatter relations. But I find the discussion here quite unconvincing, confusing the selection of a sub-ensemble by conditionalising on a random variable, with the *controllable* selection of a sub-ensemble by predictive selection. So without going into any details of the proposed experiment its purported conclusion actually contradicts a *theorem* of the formalism prohibiting super-pure states.

Popper returned to the question of thought experiments in Appendix *XII of *LSD* entitled 'On the Use and Misuse of Imaginary Experiments, Especially in Quantum Theory'. Popper warned against '*the apologetic use of imaginary experiments*' (*LSD*, p. 443, Popper's italics). Great care must be taken not to introduce idealizations or other special assumptions unless they are favourable to an opponent, or such that any opponent would have to accept them (*LSD*, p. 444). It is not clear that Popper heeded his own warnings in these matters, as will become clear as we proceed to describe the next thought experiment that he devised in order to disprove the orthodox Copenhagen interpretation of quantum mechanics.

In 1982, in the new introduction to Volume 3 of the *Postscript to the Logic of Scientific Discovery* Popper describes what he refers to as an 'extension of the EPR argument' (*QTSP*, pp. 27ff.). The idea behind the experiment was to test whether 'knowledge alone is sufficient to

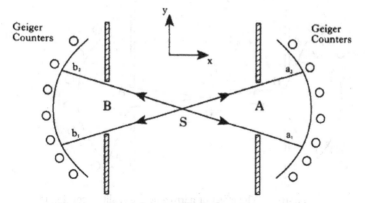

Figure 2. Popper's 1982 thought experiment.

create "uncertainty", and, with it scatter (as is contended under the Copenhagen interpretation), or whether it is the physical situation that is responsible for the scatter'.

Referring to Figure 2, Popper proposes a source S, from which pairs of particles are emitted in opposite directions in an EPR state, i.e. such that the momenta and the positions of the particles are correlated. The correlated particles are selected through slits A and B, which are sufficiently wide that there is no appreciable diffraction as they pass through the slits, and then fire off in coincidence Geiger counters arranged in a semicircle behind the slits, the counters firing between b_1 and b_2 behind the B slit and between a_1 and a_2 behind the A slit as indicated in the figure. For simplicity we just consider particles moving in the x–y plane of the diagram.

Since the momenta are correlated, b_1 fires with a_2, and b_2 with a_1, and so on. Now Popper proposes narrowing the A slit, as illustrated in Figure 3. Since the A slit is effectively constraining the y-coordinate of the particles passing through

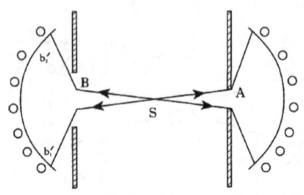

Figure 3. The effect of narrowing the A slit, according to
Popper, is to produce wide-angle scatter behind the wide slit B

it, then as a result of the position correlation, the y-coordinate
of the particle passing through B is also more tightly constrained as we narrow the A slit, and then according to the
Copenhagen interpretation, the more accurate knowledge of
the y-coordinate of the particle passing through the wide B slit
will produce a scatter in the transverse momentum, so, counters over the wider arc b_1' to b_2' will now fire. (Of course there
will be a corresponding scatter behind the narrow A slit.)

Popper is convinced that this effect would not
happen, so if performed the experiment would, according to
Popper, constitute a decisive refutation of the Copenhagen
interpretation.

The flaw in Popper's argument[3] is that he misun-
derstands the nature of the EPR correlations.

---

[3] There are other critiques of the experiment in the literature making
broadly similar points to my own. See, for example, A. Sudbury
'Popper's Variant of the EPR Experiment Does not Test the

The EPR state at the time t = o of emission of the two particles, in respect of the transverse y-dimension is of the form

$$| \psi_0 >= \int_{-\infty}^{\infty} | y > \otimes | y > dy \qquad (2)$$

$$= \int_{-\infty}^{\infty} | p_y > \otimes | -p_y > dp_y \qquad (3)$$

The form (2) demonstrates the position correlation and the form (3) the momentum correlation.

So the EPR source, according to (2) is not a point source, like S in Figures 2 and 3, but an infinite *incoherent* line source. Why incoherent? Because for any observable A_1 of particle 1, for example, the expectation value at time t is

$$< \psi_t | A_1 | \psi_t >= \int_{-\infty}^{\infty} < y_t | A_1 | y_t > dy \qquad (4)$$

where $|y_t>$ is the time-evolved state at time t which starts at t = o as an eigenstate $|y>$ of position. Similarly, for observables referring to particle 2, so expectation values are *additive*—there is no interference between Schrödinger waves originating at different values of y.

So Popper's experiment with an EPR source looks like Figures 4 and 5 showing wide angle firing of the counters behind the B slit when the A slit is wide, just the same as when it is narrow, i.e. altering the width of the

Copenhagen Interpretation', *Philosophy of Science* **52** (1985), pp. 470–76 and H. Krips, 'Popper, Propensities and Quantum Theory', *The British Journal for the Philosophy of Science* **35** (1984) pp. 253–292.

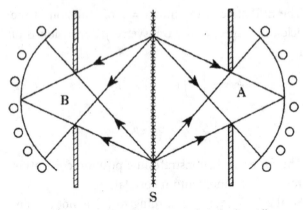

Figure 4. The EPR source with wide slits. All counters fire on both sides of the experiment

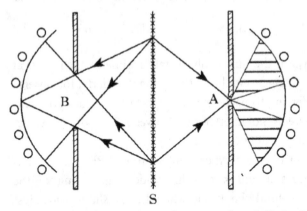

Figure 5. EPR source with a narrow slit at A. Notice diffraction effects behind A, but all counters fire on both sides of the experiment.

A slit has no effect on which counters behind the B slit fire. So the 'Popper effect' is not predicted according to the formalism of quantum mechanics, and hence not predicted according to *any* interpretation of that formalism; in particular, it is *not* predicted by the Copenhagen interpretation.

When I showed these diagrams to Popper, he refused to accept that I had a general argument that narrowing slit A could not make a counter fire behind B, that would not have fired when A was wide. Perhaps by altering the geometry he could get what he wanted. To get rid of the wide-angle firing behind B in *all* circumstances it is obviously necessary to truncate the line source. But truncating the EPR source by replacing the infinite limits in Eqn (2) by finite limits, then we lose the momentum correlation, i.e. we can no longer rewrite (2) in the form (3). In other words, with a truncated source, a narrow beam on the right in Figure 3 does not correlate with a narrow beam on the left, with resulting diffraction as Popper supposed. Figure 3 should actually be replaced by Figure 6, showing no extra diffraction behind B, as compared with the situation with both slits wide as illustrated in Figure 2.

But Popper was still not convinced! So I provided the following general proof. Consider any observable $\mathcal{A}$ for the particle passing through the A slit with eigenvalues a_2, similarly for $\mathcal{B}$ on the left with eigenvalues b_j. Then since $\mathcal{A}$ and $\mathcal{B}$ commute, we can write, in an arbitrary state $|\Psi>$

$$\mathrm{Prob}^{|\psi>}(b_j) = \sum_i \mathrm{Prob}^{|\psi>}(b_j/a_i) \times \mathrm{Prob}^{|\psi>}(a_i) \qquad (5)$$

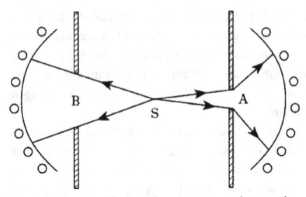

Figure 6. The loss of momentum correlation with a localized source.

Now suppose $\text{Prob}^{|\Psi>}(b_j = 0)$ and $\text{Prob}^{|\Psi>}(a_i) \neq 0$ for some particular index i. Then (5) implies that, under these conditions, $\text{Prob}^{|\Psi>}(b_j/a_i) = 0$, i.e. selecting a subensemble on the left by conditionalising on the right with an event that has a non-zero probability of occurring, cannot convert a zero probability for bi into a non-zero probability for b_j.

By a curious irony it *is* possible to produce the 'Popper effect', but not in the way Popper thought, by using a thick A slit. The effect is illustrated in Figure 7, where we suppose the thickness of the A slit is very large compared with its width.

The increased angle of firing behind the A slit is now produced by the fact that, due to diffraction effects at the narrow A slit, we can actually produce, not a subensemble, but a larger ensemble of points on the line source as compared with the case of the wide slit.

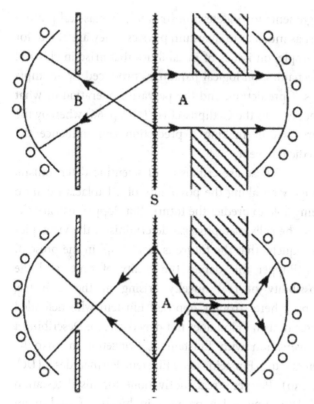

Figure 7. The Popper effect with a thick slit

2. Indeterminism and Propensities

Popper's approach to the quantum theory is based on two key ideas:

(a) The prevalence of indeterminism in physics. Instead of the slogan 'all clouds are clocks', Popper wants to substitute 'all clocks are clouds' (*OK*, p. 215). His

271

arguments for indeterminism apply in classical physics just as much as in quantum physics. They are based, for example, on issues of instabilities that arise in classical mechanics, on logical paradoxes produced by attempts at self-prediction, and the pervasive operation of what Popper calls the Oedipus-effect (*OU*, p. 66) whereby the very fact of making a prediction can influence the predicted event.

These are criticisms of scientific determinism, roughly speaking the possibility of a Laplacian demon being able to predict the future. But Popper also attacks what he calls metaphysical determinism, the view, picturesquely, that the future is 'coiled up' in the present. For Popper this denies the reality of time, and the possibility of real novelty arising in the Universe. Popper here strongly opposed Einstein who defended determinism ('God does not play dice...'). Describing a conversation with Einstein at Princeton in 1950 he refers to Einstein as Einstein-Parmenides (*UQ*, p. 130). Popper's real motivations for his detestation of determinism lay perhaps in his belief in human freedom, the unpredictability of a Mozart symphony or a Beethoven sonata, the development of new arguments and theories, the denizens of his World 3, that creative evolution is outside the purview of World 1 alone and that World 1 is crucially affected by World 3, via the mediation of World 2, and so on.

(b) For these reasons Popper thinks that probabilities in physics cannot, in general, be epistemic. How could human ignorance produce genuine physical effects?

Popper originally held a frequency view of probability, but the production of stable limiting frequencies (i.e. of random sequences) was for Popper a physical fact that needed explaining via an objective notion of tendency, and this led him in the 1950s to his propensity interpretation. This he regarded as particularly well suited to explicating the scatter relations which he used for interpreting the uncertainty relations in quantum theory.

3.   State preparation and Measurement

Popper regarded the Heisenberg uncertainty relations as limitations on the possibility of producing homogenous ensembles with precise values of position x and momentum p, for example. So Heisenberg had discovered a limitation on state preparation, not a limitation on measurement. When we measure x or p this is a backward looking exercise as contrasted with the forward looking exercise of state preparation, and such measurements are necessary to *test* the reciprocal scatter relations expressed in the Heisenberg principle.

Moreover, Popper says that he believes that between interactions particles behave classically, i.e. following Newtonian straight-line trajectories, but when interactions occur they get disturbed in an intrinsically stochastic fashion, and this is what produces the quantum-mechanical scatter.

As an *analogy* Popper often referred to what he called the pin board (i.e. a form of bagatelle—cf. e.g., QTSP, pp. 72, 87, 153). Here the balls roll deterministically between impacts with the pins. The statistical distribution of the balls at the bottom of the board is the manifestation of a propensity to produce individual results, the final position of each individual rolling ball. The propensity is a property of the whole set-up of pins and balls and is not to be thought of as a property of the individual balls themselves. This, for Popper, is the origin of what he calls 'the great quantum muddle' (*QTSP*, p. 52). The mistake here is like treating the average height of people in a room as a property of each particular person in the room. My own view is that it is perfectly legitimate to regard the quantum-mechanical propensities as properties of the particles, properties *manifested* in the context of particular experimental arrangements. This means of course that there are as many properties as there are experimental arrangements, so, following Popper, it may avoid confusion to talk of the property as belonging to the experimental arrangement, but it does not seem to me mandatory to talk in this way.

Returning to Popper and the pin board, changing the arrangement of the pins changes the propensity—indeed the arrangement of pins acts like a sort of stochastic field of force confronting the individual balls. Popper sees this as explaining, in a general sort of way, the two-slit experiment. The closing of a slit is like removing a pin from the pin board. Even if the ball on particular occasions never hits the pin, nevertheless the propensity changes. Popper, of course, stresses (in response to a critique of Feyerabend[4]) that the

---

[4] P. Feyerabend, 'On a Recent Critique of Complementarity', *Philosophy of Science* **35** (1968), pp. 309–31; **36** (1969), pp. 82–105.

pin board does not exhibit the characteristic interference phenomenon of the two-slit experiment. To explain that is a matter for the physics of the relevant propensities. Popper generally believes in the Lande approach,[5] that the effect of the two slits on the beam of electrons is related to the periodicity of the structure, i.e. the separation of the slits, which controls the transfer of momentum to the electron as it passes through *one* of the slits. If we close a slit we change the sort of object that the electron is interacting with, and hence the character of the interaction.

Turning to another aspect of measurement, the famous (or infamous) collapse of the wave function, or projection postulate, this Popper regarded as a quite unproblematic consequence of conditionalising a probability distribution on the outcome of an experiment (*QTSP*, p. 78). Consider a coin toss. When the coin lands heads, the probability of heads jumps from $\frac{1}{2}$ to 1, and the collapse of the wave function is just an instance of this type of behaviour.

Let us look at this from the point of view of joint distributions. In standard probability theory one defines conditional probabilities in terms of joint distributions, thus Prob(A/B) $\overline{\overline{DF}}$ (Prob(A&B)/Prob(B)), provided that Prob (B)$\neq$0. Popper's own approach to probability theory starts with conditional probabilities as basic and proceeds to joint distributions, essentially defining Prob(A&B/C) $\overline{\overline{DF}}$ Prob (A/B&C).Prob(B/C) (cf. *LSD*, p. 332). But either way of proceeding seems committed to joint distributions such as

---

[5] A. Landé, *Quantum Mechanics* (Pitman, London, 1951).

Prob(x&p) (perhaps conditionalized on an experimental set-up) for incompatible observables such as x and p. So what is the status of joint distribution and simultaneous measurability in quantum mechanics?

Suppose we measure an observable Q and find the value q, so the new state (according to the projection postulate) is the eigenket $|q>$. If now we measure an incompatible observable Q′ and find the value q′, then it seems reasonable to claim that in the state $|q>$, Q has the value q and Q′ the value q′. Of course we cannot use this knowledge to predict the value of Q *after* we measured Q′. This might well differ from q, due to the disturbing effect of the measurement of Q′.

But can we measure Q and Q′ when the state $|\Psi>$ is not an eigen-state of Q (or Q′)? A typical proposal here is to measure an observable, call it U, which is *compatible* with Q and has the same numerical eigenvalues as Q′, and in the state $|\Psi>$, has the same probability distribution as Q′. So U and Q′ are matched probabilistically, but it does not follow that on a particular occasion the result for U is the same as the result we would have found if we had measured Q′. The classic example of this is the time-of-flight measurement of momentum. We consider one-dimensional motion. An initial measurement of position locates a particle sharply at the origin at time t = 0, then if we measure the position X again at a later time t, the quantity U = mX/t is distributed probabilistically in the same way as the momentum scatter generated by the initial position measurement. But measuring X allows us to infer the value of U, so do the combination of values x for X and u for U (where u = mx/t) constitute a joint measurement of position and momentum for the particle. The question is, does the U measurement give the same result, on a particular occasion, as a 'direct' measurement of momentum, e.g. by bending in a magnetic field, or using a Doppler shift to measure the velocity and hence the momentum. Note that for all these direct measurements, assuming they are of the so-called first kind, so that the projection postulate applies, the results of one procedure can be checked by applying the other procedures, since they are all compatible procedures. But in the case of the time-of-flight approach, after the X measurement the momentum is disturbed, so the value *before* the X measurement cannot be checked by a subsequent direct measurement. Nevertheless, if we assume uniform rectilinear trajectories for the particle after the initial position measurement at t = 0, as Popper does, then U must be measuring the momentum during the interval *between* the two position measurements, although as we have seen this is something that cannot be checked or tested.

Now what does all this tell us about the joint distribution for x and p? Clearly at time t the joint probability density is of the form $\delta(p - mx/t) \cdot \text{Prob}(x)$, which certainly returns the right marginals for x and p. But this only works for a particular state. Indeed joint distributions can always be defined for particular states, e.g. take the joint density to be of the form $\text{Prob}(p) \cdot \text{Prob}(x)$. Again this certainly returns the right marginals. But if we ask the question, is there a quantum-mechanical observable that measures the joint quantity 'x and p' and returns the right marginals in all states the answer is 'no', according to an important theorem

of Fine.[6] Constructions in the literature[7] that do return the
right marginals in all states, do not give only non-negative
values for the joint distribution, and so these do not qualify
as *bona fide* probabilities.

So it is difficult to fit joint distributions for incompatible observables into the formalism of quantum mechanics, and indeed Arthur Fine has urged that we should reject
them, even in a phase-space reconstruction of quantum
mechanics. As Fine has argued joint distributions in probability theory only arise in general when x and p, for
example, are regarded as random variables over an underlying probability space, and we don't have to make this
overly strong assumption.[8] It would have been interesting
to know Popper's views on these questions. As far as I know
he never referred to Fine's arguments on this point.

4.   Nonlocality and Bell's Inequality

Another important issue in the philosophy of quantum
mechanics, on the kind of realist construal espoused by
Popper, is the question of proofs of nonlocality, via the
theoretically predicted and experimentally verified violation
of the famous Bell inequality.

---

[6] A. Fine, 'Probability and the Interpretation of Quantum Mechanics',
   The British Journal for the Philosophy of Science **24** (1973), pp. 1–37.

[7] See, for example, E. Wigner, 'On the Quantum Correction for
   Thermodynamic Equilibrium', *Physical Review* **40** (1932), pp. 749–59.

[8] See A. Fine, 'Logic, Probability and Quantum Theory', *Philosophy of
   Science* **35** (1968), pp. 101–11.

Popper made two sorts of comment here. Firstly, if the experiments and their interpretation were taken at face value, then he regarded the Aspect experiment, for example, as a crucial test of the Lorentz versus the Einstein interpretation of the relativity transformation equations. Instantaneous action would fix a privileged reference frame, an ether frame if you like.

But secondly, Popper also leaned to the view that the theoretical proofs of the impossibility of local hidden-variable theories for experiments such as Aspect's might contain technical flaws. He gave support to the work of Thomas Angelidis[9] who sought to demonstrate local models of these experiments. My last correspondence with Popper, just before his death, was concerned with the validity of this work. In my view the models were not correctly described as local. Popper himself admitted that the whole matter was too technical for him to check everything fully, but urged that the latest work of Angelidis should be fully replied to in the literature. This is something I propose to do on a future occasion.

At first sight it might appear that, with Popper's emphasis on indeterminism, he would be correct in

---

[9] T. Angelidis, 'Bell's Theorem: Does the Clauser–Horne Inequality Hold for all Local Theories?', *Physical Review Letters* **51** (1983), pp. 1819–22; and 'On the Problem of a Local Extension of the Quantum Formalism', *Journal of Mathematical Physics* **34** (1993), pp. 1635–53. The first paper has been criticized by a number of people. The second paper has not so far attracted critical comment in the literature.

examining the *stochastic* hidden-variable approach.[10] But in doing this he is really being inconsistent with his other views. Thus his commitment to deterministic behaviour between interactions, coupled with his adherence to a thesis of faithful measurement, means that the measurement outcomes are *deterministically* related to the state of the particles *after* they have left the source. Hence the deterministic hidden-variable proofs of the Bell inequality are the ones that need to be examined consistently with the views he expressed in other contexts.

5. Conclusion

Popper stressed (*QTSP*, p. 6) that 'the strongest reason for my own opposition to the Copenhagen interpretation lies in its claim to finality and completeness'. I entirely agree with the view that we should not rule out criticism by fiat or authority. Popper fought a lone battle against the Copenhagen interpretation at a time when anyone attempting to criticize orthodoxy was liable to be labelled at best an 'outsider' or at worst a crank. But Popper's carefully argued criticisms won the support of a number of admiring and influential physicists. He has done a great service to the philosophy of quantum mechanics by emphasising the distinction between state preparation and measurement and trying to get a clearer understanding of the true

[10] See, for example, M. L. G. Redhead, *Incompleteness, Nonlocality, and Realism* (Oxford: Clarendon Press, 1987), pp. 98ff.

significance of the uncertainty principle, but above all by spearheading the resistance to the dogmatic tranquilising philosophy of the Copenhagenists. Because some detailed arguments are flawed, this does not mean that his overall influence has not been abundantly beneficial.

9 The Uses of Karl Popper

GÜNTER WÄCHTERSHÄUSER

K arl Popper's work is of great diversity. It touches on virtually every intellectual activity. But he himself considered his philosophy of science one of his most important achievements. And indeed his achievement here is revolutionary. It destroyed the philosophy of inductivism which held sway over science for hundreds of years.

It should not surprise us that the recognition of this fact is resisted by most philosophical schoolmen. Usually the debate is carried out among philosophers. Their papers are philosophical papers and science comes in by way of interspersed examples. In this paper I shall try it the other way around, from the perspective of the scientist. I shall try to give you an account of my scientific field, the inquiry into the origin of life. And the philosophy of science will come in by way of interspersed references.

I

Before I come to my scientific story, let me briefly summarize and contrast the major tenets of inductivism and of Popper's deductivism (*LSD, RAS, BG, CR*). I begin with a caricature of inductivism in the form of eight theses:

1. Science strives for justified, proven knowledge, for certain truth.
2. All scientific inquiry begins with observations or experiments.
3. The observational or experimental data are organized into a hypothesis, which is not yet proven (context of discovery).
4. The observations or experiments are repeated many times.
5. The greater the number of successful repetitions, the higher the probability of the truth of the hypothesis (context of justification).
6. As soon as we are satisfied that we have reached certainty in that manner we lay the issue aside forever as a proven law of nature.
7. We then turn to the next observation or experiment with which we proceed in the same manner.
8. With the conjunction of all these proven theories we build the edifice of justified and certain science.

In summary, the inductivist believes that science moves from the particulars to the general and that the truth of the particular data is transmitted to the general theory.

Now let me give you a caricature of Popper's theory of deductivism, again in the form of eight theses:

1. Science strives for absolute and objective truth, but it can never reach certainty.
2. All scientific inquiry begins with a rich context of background knowledge and with the problems within this context and with metaphysical research programmes.

3. A theory, that is, a hypothetical answer to a problem, is freely invented within the metaphysical research programme; it explains the observable by the unobservable.

4. Experimentally testable consequences, daring consequences that is, are deduced from the theory and corresponding experiments are carried out to test the predictions.

5. If an experimental result comes out as predicted, it is taken as a value in itself and as an encouragement to continue with the theory, but it is not taken as an element of proof of the theory of the unobservable.

6. As soon as an experimental result comes out against the prediction and we are satisfied that it is not a blunder we decide to consider the theory falsified, but only tentatively so.

7. With this we gain a deeper understanding of our problem and proceed to invent our next hypothetical theory for solving it, which we treat again in the same way.

8. The concatenation of all these conjectures and refutations constitutes the dynamics of scientific progress, moving ever closer to the truth, but never reaching certainty.

In summary, the Popperian deductivist believes that science moves from the general to the particulars and back to the general—a process without end. Let me inject a metaphor. I might liken the Popperian view of science to that of a carriage with two horses. The experimental horse is strong, but blind. The theoretical horse can see, but it cannot pull. Only both together can bring the carriage forward. And behind it leaves a track bearing witness to the incessant struggle of trial and error.

II

We now come to my story of science. It may be broadly characterized as the development of the relationship between chemistry and biology. How is dead matter transformed into living matter?

The year 1644 marks the death of a great scientist, the Belgian physician Jan Baptist van Belmont. He had spent his life and fortunes on scientific research. Yet during his lifetime he published nearly nothing for fear of the Inquisition. In his last will he asked his son to publish his results in the form of a book: *Ortus medicinae*. It became an instant success, with translations into several vernacular languages. It culminated in a most daring thesis: 'All life is chemistry'. With this he established one of the most sweeping metaphysical research programmes in the history of science. To this day, all the life sciences, notably biochemistry, molecular biology and molecular evolution and certainly the problem of the origin of life are situated squarely within van Belmont's research programme. All the key problems in these fields come down to the problem of the relationship between animate and inanimate matter. In an ingenious approach to this vexing problem van Belmont carried out the first quantitative experiment in the history of biology and he did it quite methodically. I present his report in the English translation:

> I took an earthenware pot, placed in it 200 pounds of earth dried in an oven, soaked this with water, and planted in it a willow shoot weighing 5 pounds. After five years had passed, the tree grown therefrom weighed

169 pounds and about 3 ounces. But the earthenware pot
was constantly wet only with rain ... water.

... Finally, I again dried the earth of the pot, and it was
found to be the same 200 pounds minus about 2 ounces.
Therefore, 164 pounds of wood, bark, and root had
arisen from the water alone.[1]

Now, what is the philosophical methodology behind this
experiment? Unfortunately, the record is silent on this point.
So it would seem to be legitimate to look at this report
through the spectacles of our current philosophy of science;
in fact alternatively through inductivism and through
Popperian deductivism. We may hope for a double benefit:
(1) a clear understanding of the historical report; and (2) a
clue as to which of the two mutually exclusive philosophies
is right and which is wrong.

From the platform of our current state of knowledge
it will strike the inductivist as most important that van
Belmont's conclusion is wrong. This must mean to him that
van Belmont did not apply the proper inductive method of
science. He reports only one single experiment. There are no
repetitions. He did not repeat the test with 500 willow trees,
or with different kinds of trees, or with different kinds of
soils. One single experiment was enough for him. This
makes no sense to the inductivist. And so the inductivist
must come to view van Belmont as one of those queer,
irrational, prescientific characters, amusing but irrelevant.

[1] Cited by T. D. Brock and H. G. Schlegel, in H. G. Schlegel and B.
Bowien (eds), *Autotrophic Bacteria* (Berlin: Springer Verlag, 1989).

Now let us apply the Popperian view of science. Van Belmont was operating within a rich context of Renaissance knowledge. It was widely accepted that matter does not spring from nothing, nor disappear into nothing. And it was an equally widely held theory that the substance of growing plants comes from soil. The first theory was to van Belmont what Popper calls unproblematic background knowledge. The second theory was to him problematic and in need of testing. From both theories jointly he deduced a testable consequence. The weight gain of a growing willow tree must be equal to the weight loss of the soil in which it is rooted.

He carried out an ingenious experiment, bringing the soil before and after the growth period to the same reference state by drying. The result did not come out as predicted. 164 pounds of added tree weight compared to only 2 ounces of loss of soil weight, a small amount well within the experimental error. In the face of such a glaring result, van Belmont rightly decided that repeating such an experiment would be a waste of time and money. And so he decided to consider the soil theory falsified.

Van Helmont operated with a limited set of two possible material elements: earth and water. Having eliminated earth, the only remaining possibility was water. His result then was to him proof by elimination. This makes it understandable why he ends his report with a definitive conclusion: 'Therefore, 164 pounds of wood, bark and root had arisen from water alone.'

Today we hold that this is wrong. One of the important nutrients of plants is carbon dioxide, a gas.

Gases, however, were to van Belmont non-material spiritual entities. Therefore, by his own prejudice, he was prevented from including gases in his set of possibilities. It is ironic that it is van Belmont, who discovered that there are gases other than air, who coined the name 'gas', and who even discovered carbon dioxide.

There is an important Popperian lesson to be learned here. In science our sets of possible solutions should never be taken as exhaustive. They are limited by our limited imagination and by our more or less unconscious prejudices. Inductivist philosophers have always missed that simple point.

Here we have now a stark difference between inductivism and Popper's deductivism. The inductivists lead us to view science as a gigantic book-keeping affair and major parts of the history of science as irrelevant or even ridiculous. By Popper's account the same history of science is seen as a fascinating intellectual adventure story and instead of heaping ridicule on our scientific forebears we see them as the giants they are.

III

After van Helmont's famous thesis: 'All life is chemistry', and the recognition that plants feed on carbon dioxide and light while animals feed on plants, it became accepted belief that plant chemistry and animal chemistry were deeply divided, as deeply as both were divided from mineral chemistry. In this situation, in the year 1806, Jons Jacob Berzelius, a Swedish chemist, came out with two bold conjectures. He held that there was an essential unity between plant and animal chemistry which he came to call 'organic chemistry'. This he distinguished from 'inorganic chemistry'. But more importantly, he held that there was an unbridgeable gap between organic and inorganic chemistry. His central dogma can be formulated as follows:

> The generation of organic compounds from inorganic
> compounds *in vitro*, outside a living organism,
> is impossible.

He believed that a special force was at work inside all living beings which he called '*vis vitalis*'.

The year 1828 marks a watershed in the relationship between chemistry and biology. Friedrich Wohler published a simple experiment. He reacted two wholly inorganic compounds, ammonium chloride and silver cyanate, and he produced urea, a compound which had only been found in the urine of animals. Wohler wrote triumphantly: I can make urea, and don't need a dog for this. He knew, with one single falsifying experiment, he had written himself into the annals of chemistry. There is not a shred of inductivism in this story. In fact Justus von Liebig, Wohler's contemporary and friend, wrote a whole book to free science from the plague of inductivism.[2] But against his burning protest the nineteenth century sees the relentless spreading of this plague.

But let us go on with our story. The nineteenth century had already acquired a rich chemical picture of the

---

[2] J. v. Liebig, *Induktion und Deduktion* (1865).

world. Living organisms synthesized their constituents *in vivo* from inorganic matter, and chemists were able to synthesize these same constituents *in vitro* and also from inorganic matter. Against this backdrop the next major question came into scientific focus. Aside from the obvious reproduction of higher organisms, where do living organisms come from in the first place?

There was an intuitively obvious answer, which had come down through the ages. The simpler organisms, the insects, the worms, the bacteria arise by spontaneous generation from decaying dead organic matter. In the year 1861 the French microbiologist Louis Pasteur published an ingenious experiment. He used two identical bottles with the now famous Swan-neck. He filled both with a sterilized nutrient broth. The first bottle was kept upright so that bacteria in the air would not enter. The second was tipped so that the bacteria could come in contact with the broth. The first stayed perfectly free of bacteria while the second quickly developed a dense growth. This was a resounding refutation. The theory of spontaneous generation of living organisms was laid to rest.

Now microbiology had its central dogma:

> The generation of whole living organisms from chemical compounds, outside a living organism is impossible. Life can only spring from life.

Again, there was no inductivism here. There was merely a refutation of an alternative.

In the year 1859 Charles Darwin published his daring hypothesis, that all organisms are the evolutionary descendents of a common ancestor.[3] This led automatically to the question of the origin of this primordial ancestor. In the year 1871 Darwin himself gave an answer to this question. He wrote in a letter

> It is often said that all the conditions for the first production of a living organism are now present, which could ever have been present. But if (and oh! what a big if!) we could conceive in some warm little pond, with all sorts of ammonia and phosphoric salts, light, heat, electricity, etc. present, that a protein compound was formed, ready to undergo still more complex changes, at the present day such matter would be instantly devoured or absorbed, which would not have been the case before living creatures were formed.[4]

Eight years earlier the German biologist Mathias Jacob Schleiden, one of the founders of the cell theory, had suggested that a first cell might have been formed under the entirely different atmospheric conditions of the young earth.[5]

Neither Darwin's nor Schleiden's proposals can be said to have great scientific value. They were not concrete enough to have explanatory or predictive power. They were

---

[3] C. Darwin, *On the origin of species by means of natural selection of the preservation of favoured races in the struggle for life* (London: Murray, 1859).

[4] F. Darwin, *The life and letters of Charles Darwin* (London: Murray, 1887), Vol. III, 18.

[5] M. J. Schleiden. *Das Alter des Menschengeschlechts: die Entstehung der Arten und die Stellung des Menschen in der Natur—drei Vortriige fiir gebildete Laien.* (Leipzig: Engelmann, 1863).

components of a very vague metaphysical research programme.

IV

In the early 1920s the Communist party of the Soviet Union came to the conclusion that its atheistic campaign would be bolstered if it could be shown scientifically that the origin of life did not require a divine intervention. And so, by the account of Christian de Duve,[6] it was Alexandro Iwanowitsch Oparin, the biochemist, and later Lyssenkoist who received the order from the party to produce such a theory. It was produced and published in 1924.[7] This theory incorporated the suggestions of Darwin and Schleiden. By Popper's standards Oparin's theory should have departed from these vague proposals by going in the direction of greater concreteness. This would have generated explanatory power. That means the power to explain many facts of today's organisms with few assumptions. And it would have generated predictive power, which means testable, falsifiable consequences.

But this is precisely what did not happen. Oparin's primary impetus was political. He strove for convincing power. And so he designed his theory to be immune to criticism or falsification. He invented the so-called 'prebiotic

---

[6] Ch. de Duve. *Ursprung des Lebens* (Heidelberg: Spektrum Akademischer Verlag, 1994).

[7] A. I. Oparin. *Proiskhozhdenie zhizny* (Moscow: lzd. Mosk. Rabochii, 1924).

broth', but its contents were left completely vague. This basic flaw was not corrected by the other men who published early papers after Oparin, notably the Marxists J. B. S. Haldane[8] and J. D. Bernal.[9] The theory remained vague and untestable, and so it remained untested for thirty years.

The situation changed decisively when the American chemist Harold C. Urey published his theory that the primordial atmosphere of the earth consisted mainly of methane and ammonia.[10] In the same paper he proposed that the prebiotic broth was stocked with the compounds which are formed when lightning strikes such a primordial atmosphere. Now, for the first time, a portion of Oparin's theory of a prebiotic broth was testable. Stanley L. Miller, a student of Urey, carried out the test: electric discharges in an atmosphere of methane and ammonia above water. He produced mostly a brown tar, but also, amazingly, small amounts of amino acids.[11]

Now the science community made what some may call an inductive inference. From something experimental and observable they inferred something historical and unobservable: The prebiotic broth contained amino acids. Amino acids alone cannot make an organism. Many other components are needed: purines, pyrimidines, sugars, lipids, tetrapyrrols and coenzymes. In the next forty years numerous

---

[8] J. B. S. Haldane, *Rationalist Ann.*, **3** (1929).

[9] J. D. Bernal, *Proc. Phys. Soc. (London) Sect. A*, **62** (1949), pp. 537–558.

[10] H. Urey. *The Planets: Their Origin and Development* (New Haven: Yale University Press, 1952).

[11] S. L. Miller, *Science*, **117**, (1953), pp. 528–529.

Miller-style experiments were carried out. The conditions were modified with the aim of generating some of these other components. As soon as traces of another compound were found, this too was claimed to have been in the pre-biotic broth. In this way the broth became stronger and stronger but the theory became weaker and weaker. In 1982 the German physicist Manfred Eigen proclaimed that he had no doubt that the prebiotic broth contained all kinds of biomolecules and that it was something like a nourishing beef tea.[12] And about five years later the American Alan M. Weiner wrote in a standard textbook of molecular biology:

> Indeed, it would not be an exaggeration to say that every expert in the field of molecular evolution has a different notion of what exactly was in the prebiotic soup.[13]

The situation grew still worse. Most of the supposed 'pre-biotic' reactions require chemical conditions which are incompatible with the conditions of most of the others. Therefore, it was concluded that there must have been several separate cauldrons with prebiotic broth, with differ-ent chemical conditions. Others claimed that the soup was significantly enriched with meteor material or unknown cometary material or unknown ingredients from interstellar

[12] M. Eigen, W. Gardiner, P. Schuster and R. Winkler-Oswatitsch, *Scientific American*, **244** (1981), pp. 88–118.

[13] A. Weiner, in J. D. Watson, N. H. Hopkins, J. W. Roberts, J. A. Steitz and A. M. Weiner (eds), *Molecular Biology of the Gene* (Menlo Park: The Benjamin/Cummings Publishing Company, Inc., 1987), pp. 1098–1163.

dust grains. Others speculated that the prebiotic broth would run out of one lake, and on to hot volcanic rocks and then with more rain into another lake. If we check this situation against Popper's theory of science we can make two observations:

1. With every modification, the prebiotic broth theory increased in vagueness and ambiguity and it decreased in falsifiability and it decreased in explanatory power. The development of the theory is counter-scientific.

2. Most workers in the field were inductivists. They believed that the sum total of the experimental results would tell us all about the prebiotic broth. There is perhaps no other example in the whole of science which violates the principles of Popper's theory of science more thoroughly. And there seems to be no other scientific enterprise, which has suffered a similar devastation from the attitudes of justificationism and inductivism as the prebiotic broth theory. It is a perfect example for the consequences of a continued application of the wrong methodology of science.

V

Let us now apply the Popperian methodology. We recognize at once: the problem is not how we can strengthen atheism. Our problem is one of explanation. How can we explain the mountain of facts of biochemistry. Now we see the magnitude of the problem. We have to explain biological facts which exist today with a chain of evolutionary events which stretches over four billion years. Our explanatory problem is clearly complex.

Now, many, scientists have been very successful without ever thinking about the theory of knowledge. Others have been philosophical inductivists, but in their scientific practice—quite inconsistently—deductivists. In the face of our complex problem we cannot hope to succeed with either attitude. I propose, that we can succeed only if we consciously and consistently apply a methodology of science and only if this methodology is not fundamentally flawed.

Our problem is biological. The solution will be a theory of biology, a theory about the overall process of biological evolution. But let us be clear about our metaphysical outlook. We consider the process of evolution a historic process. If in a thought experiment we would start it again under exactly identical conditions we would expect it to run quite a different course. And if started again, another different course. This is because we consider that at any point in this long process the number of possibilities exceeds by far the number of simultaneous actualizations. This is what we mean when we consider the process of evolution contingent and indeterministic.

If we could trace this historic process backwards, we would expect to end up in purely chemical processes. But the theories of chemistry are universal, independent of space and time. Here we see the next difficulty: our desired comprehensive theory of evolution must trace a unique historic process of biological evolution into a universal process of chemistry. This means that we have to aim at a universal theory of evolution independent of the particular chemical situations on the planet earth and notably independent of special assumptions about nucleic acids or the like.

Next we have to consider the metaphysical problem of determinism. Most physicists agree with Karl Popper that the laws of physics are indeterministic in the sense that we cannot predict the fate of individual particles, atoms, molecules with precision. The process of evolution is based on singular events of mutation in the replication of singular DNA molecules in singular cells. They are therefore indeterministic in the sense of physics. Chemistry, however, is a science that is concerned not with single molecules but with huge ensembles of molecules. The experimental results are statistical results and in this sense of course predictable. If we consider additionally that evolution proceeds in the direction of increasing complexity, we arrive at the following overall picture of evolution. Think of a landscape of chemical possibilities. The first form of reproduction and the earliest phases of evolution may occur in a narrow canyon, perhaps with unique singular possibilities. As the complexity increases the number of chemical possibilities increases. But all possibilities are actualized. Only after a certain degree of complexity has been reached will the number of possibilities begin to exceed the number of actualizations.

This means that the whole picture of self-organization may well be mistaken. Instead we might have to evoke the picture of a self-liberation process, a process which creates its own prospects—an unfolding of possibilities; a process that begins in necessity and ends in chance. In the earliest phases of this overall process of evolution, a biological selection is not required. It comes into the picture only later to keep the explosion of possibilities at bay. This is how the interface between the historic process of evolution and the universal laws of chemistry may well turn out to be.

How shall we proceed in building a more concrete theory? Karl Popper spent a lifetime fighting against reductionism; against the metaphysical notion that biology can be reduced to chemistry. If we adopted for a moment the reductionist position we would try to derive the desired universal theory of evolution from first chemical principles, say from the differential equations of quantum chemistry and chemical initial conditions. Nobody has ever seriously entertained such a reductionist position. We must expect that the landscape of chemical possibilities has huge unknown continents. The chemist will not have the solution to our problem.

This means we have to turn to biology. We have to begin with today's organisms and try—hypothetically—to follow the river of evolution upstream—backwards in time; with the hope of arriving at its source. If we do this, we do it with a certain hope. That all organisms have highly conserved features which tell us about our distant past. The earth is 4.6 billion years old. The oldest microfossils are 3.5 billion years old. The oldest sedimentary rocks are 3.8 billion years old. But life on earth must be still older. So this is now our hope: That the conserved features in the organisms living today are older than the most ancient rocks.

What was the earliest organism like? At this point many philosophers and scientists tend to fall into a trap, into the trap of essentialism—of definitionalizing. They expect to gain real wisdom by finding a definition of the word 'life'. Popper spent a lifetime fighting against such essentialism. So we will avoid this trap. We will try instead to elucidate the process of evolution and treat the problem of naming as secondary.

I now introduce two hypothetical postulates of my general theory of evolution:[14,15]

(1) All processes of biological evolution are based on a process of reproduction: An entity takes up food, grows and divides into two entities that take up food.
(2) Variations occur due to by-products with a dual catalytic feedback effect: with an altruistic feedback and an egoistic feedback.

By the altruistic feedback the catalyst promotes the reproduction process from which it is derived. By the egoistic feedback the catalyst promotes its own formation. An altruistic feedback alone is not inheritable. An egoistic feedback alone is destructive. Only both jointly constitute evolution.

At this point I am still operating in a Popperian metaphysical research programme. As a scientist I have to get much more concrete. We have to ask a particular question: How did the particular process of evolution on the planet earth begin and how did it go on? Our inquiry may be divided into two phases which may partly overlap. The

[14] G. Wächtershäuser, *Microbiol. Rev.*, **52**, (1988), pp. 452–484.
[15] G. Wächtershäuser, *Progr. Biophys. Mol. Biol.*, **58** (1992), pp. 85–201.

first phase is one of a strictly theoretical inquiry. The second phase is the experimental phase.

How can we make progress with a theory without experiments? Popper has given the answer. We orient ourselves on the principle of relative explanatory power. And of course we check our modifications constantly against the backdrop of unproblematic theories of physics, chemistry, geology, etc. And now comes a most important point. With every step in the ascent of our theory we are confronted with a plurality of possibilities. Of course we should try to formulate them all. But, of course, we will not succeed. If our imagination is very limited and we think of only one single possibility, then we might have the illusion that our task is the task of proving this result of our poor imagination. But if we are lucky enough to think of several alternative possibilities, then we recognize instantly, that our task is the task of elimination. If we are lucky, then we can eliminate all but one possibility, or one set 'of possibilities'. We then proceed to our next problem which we treat in the same manner. The trick in such procedure is this: We should begin with those problems for which our theoretical elimination process promises to be least ambiguous.

In this fashion I confronted the following questions.

- What was the first food?
- What was the first energy source?
- What was the first autocatalytic reproduction cycle?
- What was the first form of division?
- What was the first form of structural coherence?

For each of these questions I tried to steer the process of elimination so that the explanatory power would be maximized. Each answer was chosen such that it explained not only one, but several facts of today's biochemistry.

My present, tentative set of answers may be summarized in simplified form.

- The first food for life was carbon dioxide.
- The first energy source was the formation of pyrite from iron sulphide and hydrogen sulphide.
- The first autocatalytic cycle was an archaic version of the reductive citric acid cycle.
- The first form of division was the cleavage of a large, unstable molecule into two molecules.
- The first form of structural coherence was the bonding of the constituents onto the pyrite surface.

And now comes an interesting observation. My theory instantly found many supporters, among them great scientists. They all said that they were not convinced of its truth, but they liked it for its explanatory power. In this fashion my theory had a peculiar degree of success before the first experimental shot was fired.

The experimental programme is now coming into full swing. With each experimental problem you are facing a huge variability of the parameters. So the function of the experiments is again mainly that of elimination. You always have expectations of course. But any positive result *vis-à-vis* such an expectation is situated within a field of negative results—if the experiments are set up with the right attitude; an attitude based on the understanding of the poverty of our imagination. And if we run into a situation, where all results are negative, we have to modify the theory again or we stop since we are at our wits' end.

I call my theory of the early evolution of life on earth the 'Iron-Sulphur-World', for iron-sulphide is part of the energy source, the battery, that is seen as driving the whole press. But is this the only possible world of life? Could there be a cobalt-sulphur world, or a nickel-sulphur world, or an iron-selenium world? The Universe is a huge place. But from a chemical point of view it is quite monotonous. This is what our theories tell us. The same some ninety stable elements will be found anywhere. And they are all formed by the same few nuclear processes. And by the laws of these processes—as our theories see them—the proportions of the elements would be similar everywhere. For example where there is nickel for a possible nickel-sulphur world, there will always be a predominance of iron.

This gives us a peculiar cosmological outlook. Throughout the universe, life might well have basically only one way of getting started. And wherever the conditions are right, it would start continuously—anywhere, anytime in the same unique way.

With this highly speculative cosmological view, our whole problem situation appears to shift. There are many different chemical spaces. In perhaps only one of them— perhaps very narrowly restricted—we expect to find the original homestead of life. Here the starting process will run continuously. So the origin is not a time—it is a place. And the process of evolution is seen as a process of conquering ever new spaces; it is seen primarily as a spatial affair; and time comes in by way of the history of the conquering of space.

Metaphorically speaking, the process of evolution is a process of liberation—of liberation from the narrower chemical confines of an iron—sulphur world and from a two-dimensional existence on pyrite surfaces. This process of liberation has now been going on for some four billion years. And it is still going on. But only at a price; at the price of unfathomable complications and ever more sophisticated controls.

10   Popper and Darwinism

JOHN WATKINS

I

The first Darwin Lecture was given in 1977 by Karl Popper. He there said that he had known Darwin's face and name 'for as long as I can remember' ('NSEM' p. 339); for his father's library contained a portrait of Darwin and translations of most of Darwin's works ('IA', p. 6). But it was not until Popper was in his late fifties that Darwin begin to figure importantly in his writings, and he was nearly seventy when he adopted from Donald Campbell the term 'evolutionary epistemology' as a name for his theory of the growth of knowledge (*OK*, p. 67). There were people who saw evolutionary epistemology as a major new turn in Popper's philosophy.[1] I do not share that view. On the other hand, there is a piece from this evolutionist period which I regard as a real nugget.

I call it The Spearhead Model of evolutionary development. It appeared briefly in the Herbert Spencer lecture he gave in 1961, which he wrote in a hurry and left in a

---

[1] I am thinking especially of the late Bill Bartley; see G. Radnitzky & W. W. Bartley III (eds), *Evolutionary Epistemology, Theory of Rationality, and the Sociology of Knowledge* (Open Court, 1987), part I.

rough and unready state. It contained mistakes that would, and did, dismay professional evolutionists. Peter Medawar advised him not to publish it,[2] and it lay around unpublished for over a decade. He eventually published it, with additions but otherwise unrevised, in Chapter 7 of *Objective Knowledge*. It did not, so far as I know, evoke any public comment from biologists or evolutionists.

When I discussed the neglect of Popper's Herbert Spencer lecture with Bill Bartley, in 1978, he was pretty dismissive, saying that it was all in Alister Hardy. Now it is true that much of that lecture is about an idea, which Popper was later to call 'active Darwinism ('PMN', pp. 39f.), which had indeed been anticipated by Hardy. Hardy had suggested that an animal's interests might change in such a way that certain bodily mutations that would previously have been unfavourable now become favourable; and he added that while such a change of interest might be forced upon the animal by external circumstances, it might result from exploratory curiosity and the discovery of new ways of life. He gave the example of forebears of the modern woodpecker switching their attention from insects in the open to insects in the bark of trees. No doubt these proto-woodpeckers proceeded rather clumsily at first; but going after this rich new food supply with tools not well adapted to the task proved at least marginally more rewarding than persisting with old habits. So new habits developed; and mutations that made the bird's bodily structure better adapted to these new

[2] He is the expert mentioned by Popper on p. 281 of *OK*.

functions now became advantageous, though they would previously have been disadvantageous. So new shapes of claw, beak, tongue, etc. began to evolve. Popper put forward the same idea, together with the woodpecker example, without reference to Hardy. But I am sure that Popper was entirely innocent here. He gave this lecture in 1961. Hardy presented the idea in his book *The Living Stream* in 1965. Before that he had broached it in the Linnean Society's *Proceedings* in 1957, but Popper would not have seen that.[3] If Popper made a mistake it was to publish this lecture essentially unrevised in 1972. When he sent Hardy a copy of *Objective Knowledge* he apologised for its lack of any reference to him.[4] He made ample amends later in 'RSR', *SB* and 'PMN'.

In any case the Spearhead Model is distinct from 'active Darwinism'. It is about certain relations between an animal's central control system and its motor system. Popper was usually a good publicist for his own ideas, and he subsequently gave plenty of publicity to the idea of 'active Darwinism'. But he allowed the Spearhead Model to fall into neglect. After he read a brief discussion of it in my contribution to the Schilpp volume on Popper, he told me that he'd forgotten it! I will resurrect it. In case you are wondering whether a neglected contribution to evolutionary theory is a suitable subject for a philosophical paper, I may

---

[3] See A. Hardy, *The Living Stream* (Collins, 1965), and *Proceedings of the Linnean Society*, **168** (1957), pp. 85–7.

[4] I am here drawing on the Popper papers.

add that the Spearhead Model has important implications, which I will try to spell out, for the mind–body problem.

But first I will say a few words about the relation between Darwinism and Popper's theory of knowledge. Against my contention that Darwin's ideas do not seem to have had a serious impact on Popper before about 1960 it might be objected that their influence had shown itself already in *Logik der Forschung* (1934); for Popper there declared that the aim of the method of science is to select among competing hypotheses 'the one which is by comparison the fittest, by exposing them all to the fiercest struggle for survival' (*LSD*, p. 42). And I agree that there is a partial analogy between his conception of scientific progress through conjectures and refutations and Darwin's conception of evolution through variation and natural selection. But there are also serious disanalogies. Perhaps the main one is this. According to Darwin, any *large* variation is sure to be unfavourable; to have any chance of being favourable, a variation has to be very slight. And this of course means that evolutionary developments are gradual and slow. But science during the last four centuries has been evolving, if that's the word, by leaps and bounds. *Inductivism* may view scientific progress as a smooth, incremental process, but Popper's view of it is essentially saltationist, the new scientific theory usually conflicting radically with its predecessors at the theoretical level and making small changes at the empirical level. For Darwin there could be no such thing as a 'hopeful monster'; but the history of science, seen through Popperian eyes, is full of 'hopeful monsters'. As I see it, the relation of a revolutionary new scientific theory to its predecessors is not unlike that of a jet-engined aircraft to its propeller-driven predecessors; the new theory takes over the work done by them, and does it better, being driven by a more powerful theoretical ontology. But as Richard Dawkins pointed out, for a jet-engined aircraft to have evolved from a propeller-driven one in a *Darwinian* manner, each nut, rivet and other small component of the earlier plane would have had to change, one at a time and by series of very small steps, into a component of the later one.[5]

I do not find Popper paying much heed to Darwinism in his middle years. There are no significant references to Darwin or Darwinism in *The Open Society* (1945) and no references in *Conjectures and Refutations* (1963). There is a brief discussion of Darwinism in 'The Poverty of Historicism' (1944–5), but its tendency is rather deflationary. Understandably anxious to dispel the idea that evolutionary theory gives any support to the historicist thesis that society is subject to a law of evolution, he said that it has 'the character of a particular (singular or specific) historical statement. (It is of the same status as the historical statement: "Charles Darwin and Francis Galton had a common grandfather".)' And he endorsed Canon Raven's dismissal of the clash between Darwinism and Christianity as 'a storm in a Victorian tea-cup' (*PH*, pp. 106–107 and n.).

But by the time he was writing his intellectual autobiography for the Schilpp volume, around 1968–9,

---

[5] See Richard Dawkins, *The Extended Phenotype* (Oxford University Press, 1982), pp. 38–39.

the situation had changed. He devoted a whole section of it to Darwinism, whose great importance he now proclaimed. The suggestion that it has a merely historical character was silently abandoned; but Popper vacillated considerably over what character it does have. He said that Darwinism is 'almost tautological' ('IA', p. 134), which doesn't sound too good, coming from someone who extols a high falsifiable content in scientific theories. He also said that Darwinism is a metaphysical research programme. Does that mean that it makes no predictions? Well, he conceded that it predicts the *gradualness* of all evolutionary developments, but he added that this is its *only* prediction. But there he was surely wrong. If a species gets geographically split in two, for instance by geological changes, and there are considerable ecological differences between the two geographical areas, Darwin's theory predicts that, provided neither population becomes extinct, there will in due course be two species between which there can be no interbreeding.

Having declared it to be a metaphysical research programme, Popper added that Darwinism is not 'just one metaphysical research programme among others' ('IA', p. 135). Then what singles it out? In answering this Popper went off on a new tack; and what he now said made it sound as though it was not so much that his theory had benefited from Darwinism, but that Darwinism could now benefit from his theory—or at least that the benefit had been mutual. He wrote: 'I also regard Darwinism as an application of what I call "situational logic".' And he added: 'Should the view of Darwinian theory as situational logic be acceptable, then we could explain the strange similarity between my theory of the growth of knowledge and Darwinism: both would be cases of situational logic' ('IA', p. 135).

I find this baffling. What he had called 'situational logic' involved an agent in a well-defined situation, for instance a buyer in a market, where the agent's situational appraisal and preferences jointly prescribe a definite course of action. What has that to do with the theory of evolution? There is no inconsistency in supposing both that all creatures always behave in accordance with the logic of their situation, *and* that all species are descended unchanged from original prototypes. Situational logic has nothing to say about the two assumptions that differentiate Darwin's theory from contemporary alternatives to it, namely that heritable variations occur and that a successful variation may get preserved.

Nor does situational logic say anything about an assumption that differentiates Popper's theory of the growth of scientific knowledge from Humean and other empiricist views, namely that science essentially involves *intellectual innovation*. Indeed, situational logic and intellectual innovation are inimical to one another. In October 1948 Bertrand Russell found the flying-boat in which he had landed at Trondheim starting to sink. When he got outside, still clutching his attaché case, he found that there was a rescue boat about 20 yards away which, for safety reasons, would not come any closer. He did not attempt to bring new ideas to bear on his problem, but acted in accordance with the logic of his situation, which required that he throw away his attaché case and swim those 20 yards. But when, years earlier, he had been struggling innovatively with the paradoxes he had discovered at the foundations of mathematics, he had no situational logic to prescribe a course for him; nor did situational logic provide a compass for Newton when he was 'voyaging through strange seas of thought alone'.

I don't think that Popper ever came up with a satisfactory answer to the question, 'Why is Darwinism important?'

II

I turn now to the nugget which I discern in Popper's later, evolutionist period. Although it needs revamping, it is potentially a significant contribution to Darwinian theory, enhancing the latter's problem-solving power. and I will begin by indicating a problem which the Spearhead Model helps to solve.

We have already met Darwin's insistence that only very slight variations have any chance of being favourable. Both T. H. Huxley and A. R. Wallace objected to this. Chapter 6 of Wallace's *Darwinism* (1889) is entitled 'Difficulties and Objections'; and its first sub-heading is: 'Difficulty as to smallness of variations'. He there said that Darwin had exaggerated how slight a favourable variation has to be, thereby inviting the objection 'that such small and slight variations could be of no real use' (p. 127). Was Wallace right? R. A. Fisher calculated that 'a mutation conferring an advantage of 1 per cent in survival has itself a chance of about 1 in 50 of establishing itself and sweeping over the entire species'.[6] But a *very* slight variation would be more likely to confer an advantage of the order of 0.001 per cent or perhaps 0.0000001 per cent; what chance would it have of establishing itself? Darwin's gradualism seems to create the difficulty that variations will be either too large to be favourable or too slight to catch on. The Spearhead Model can be of assistance here, as we will see.

The Spearhead Model involves a dualism between central control system and controlled motor system. Think of an animal facing an urgent survival problem—a leopard-ess, for instance, with hungry cubs to feed. She sets out and in due course spots an impala; does her retinal image of this desirable object stimulate this powerful motor system of limbs, claws, teeth, etc. to go bounding after it? Certainly not; the motor system is under skilful and precise central control. She moves silently, keeping upwind and out of sight. When she eventually springs, she anticipates the impala's escaping movement. Although hungry, she does not at this stage feed off the carcass, but drags it back to the lair where her cubs are waiting.

According to the Spearhead Model, motor systems and control systems are genetically independent of one another. Of course, it might happen that a mutation that brings about a change in one also brings about a change in the other, but it would be an incredible coincidence if it brought about *co-ordinated* changes. And the model's main

[6] R. A. Fisher, *The Genetical Theory of Natural Selection*, (Clarendon Press, 1930), pp. 77–78.

message is that in the evolution of species, *developments in control systems pave the way.*

The main objection to the claim that a large variation may be favourable is, I suppose, that the organ in which it occurs will thereby become maladjusted with respect to other organs with which it had been in equilibrium. More specifically, an organ whose size or power is enhanced by a large variation would make overtaxing demands on the rest of the system; a considerable lengthening of a wing, say, might overtax the wing muscles, and a considerable strengthening of wing muscle might overtax the heart. I now introduce a point which Popper did not make explicit. It is that this objection falls away when we turn from the animal's motor system to its control system. Considerable enhancements *there*, I assume, do not have an overtaxing effect elsewhere. Giving a computer more powerful software, even improving the hardware itself, say by adding a new chip, puts no extra demand on the electricity supply. And I will assume that much the same holds for animals' control systems. Carrying around an over-sized wing would be costly, but carrying around even a lot of surplus control capacity may be virtually cost-free.

Popper presented a thought-experiment which begins by envisaging a state, call it state (i), in which an animal's control system and motor system are, in his words, 'in exact balance' (*OK*, p. 278); by this he seems to have meant that the control system can cope, but only just, with the motor system. Let C and M denote respectively the control capacity and the motor power of the two systems when they are in this exact balance. Popper next envisaged two alternative developments: into a state (ii) in which control capacity C remains constant and motor power M receives an increment, and into a state (iii) in which M remains constant and C receives an increment. In the diagram the thin lines stand for control capacity and the thick ones for motor power:

C	C	C+ΔC
M	M+ΔM	M
(i)	(ii)	(iii)

In (ii), where motor power exceeds control capacity, the result, Popper declared, would be *lethal*. What about (iii), where control capacity exceeds motor power? As Popper saw it, there would be no immediate advantage; but it might be *extremely favourable* (*OK*, p. 278) later, if motor power were suitably augmented. We may call these three states respectively 'balanced', 'overpowered' and 'underpowered'.

I will refer to this as the Spearhead Model Mk I. As it stands, it is exposed to several objections. Perhaps the main one is this. We are here interested only in very slight variations, at least on the motor side, on the Darwinian assumption that only these have any chance of being favourable. There can of course be no objection to the idea that the same slight variation may be unfavourable in one context and favourable in another; but that it should prove *lethal* in one context and *extremely favourable* in another seems far-fetched. Here is a simple thought-experiment: we start with a system in an under-powered state (iii), with control capacity exceeding motor power quite considerably. Control capacity now remains constant for a period during which there is a sequence of 100 small increments of motor power, the 99th increment bringing the system back into a new balanced state (i), and the 100th tipping it over into a new over-powered state (ii). On Popper's account all these small increments will be favourable except for the last, which will be lethal. Continuity considerations argue against such a sensational reversal.

The source of the trouble, I believe, was Popper's implicit use of an all-or-nothing notion of control, whereby the motor system is either fully under control or else, the moment its power rises above a critical level, quite out of control. That is what generates the abrupt switch from favourable to lethal; and it has the unwanted implication that, given a motor system that is fully under control, no enhancement of control capacity will bring any advantage so long as the motor power remains constant. Then why should mutations get preserved which merely endow the system with potential control over motor power it doesn't yet have? Popper side-stepped this question. One of his sentences began: 'Now once a mutation like this is established ...' (*OK*, p. 278), but *how* it might get established he did not say. He allowed that mutations of this kind are 'only indirectly favourable', but claimed, as we saw, that once (somehow) established they may prove *extremely* favourable. But natural selection lacks foresight; a gene that is not now favourable will not now be selected for, however favourable it would prove later if it were selected now. One begins to see why evolutionists did not find this model persuasive.

The suggestion that a motor system abruptly switches from being fully under control to being quite out of control the moment its power rises above a critical level presupposes that its power is always used to the full. But surely power can be exploited judiciously. When her Mini won't start, a wife borrows her husband's Jaguar to go shopping. This car would soon be out of control if she or any other driver kept its accelerator fully depressed; but on her shopping expedition she is content to draw on only a fraction of its reserves of power. However, there is a countervailing consideration here. Imagine that this Jaguar is now being used as a getaway car with a police-car in hot pursuit; its driver may use its large reserves of power too freely, perhaps with lethal results. (Within a few weeks of first putting that thought on paper I read in the newspaper of five or six cases where a criminal was either killed or injured when his getaway car crashed.) While properly sceptical of the idea that the system will get hopelessly out of control the moment its power rises the least bit above a critical level, we can agree that the risk of its getting out of control increases the more its power exceeds its control capacity.

In an attempt to revise this model so that it complies with continuity requirements, I will now present a thought-experiment which relates variable motor power to variable control capacity. It involves racing cars of varying power, and drivers of varying skill. Imagine a fleet of 100 racing cars whose engine power ranges from immensely powerful down to that of, say, a minicab. The cars are otherwise as similar to one another as is consistent with this very varying motor power. To drive them there is a corps of 100 drivers whose racing skills likewise range from that of a reigning Grand Prix champion down to that of, say, a minicab driver. The racing is run by a central management, which organizes tournaments in the course of which the drivers are repeatedly reshuffled among the cars. The route, which is varied from race to race, is picked out from a complicated road network, and always includes some quite long straight stretches that end in hairpin bends. In the interest of safety the physical road is wide, but it has a rather narrow centre lane, painted white, on either side of which are grey areas, with black areas beyond them. Cars are supposed to stay in the centre lane. Straying into a grey zone, the equivalent of getting partially out of control, incurs a stiff penalty, while straying into a black zone, the equivalent of getting right out of control, incurs disqualification. The cars are clocked as they go round the circuit, one at a time, and any straying into grey or black zones is automatically recorded. All drivers have a strong incentive to perform well, an ideal performance being to complete the circuit in the shortest time and without penalties. I assume that, as well as a metric for the cars' motor powers or M-values, there is a metric, analogous to chess masters' Elo-ratings, for the drivers' racing skills or C-ratings. The combination of a certain C-rating with a certain M-value will determine a level of *racing fitness*, as I will call it in analogy with the notion of biological fitness.

Let us now pick on a driver with a C-rating in the middle range, and ask whether for him there is a car with an optimum M-value, that is, one which maximizes his racing fitness. There will surely be cars whose M-values are too low for him. What about cars at the top end of the range? Would he do best in the car with the highest M-value of all? I say that he would not. Consider what would be likely to happen when he gets to one of those straight stretches. Being strongly motivated to win, he will want to go flat out; but unless he judges it nicely he will be liable to get into difficulties at the next bend. In short, there is a danger that he will resemble the desperate driver of a getaway car. Now if there are M-values that are too high for him, and ones that are too low, the principle of continuity tells us that somewhere in between there is a value that is neither too low nor too high but *optimum*.

In Figure 1 variable control skill C is represented along the vertical axis and variable motor power M along the horizontal axis. Each isobar depicts the common level of

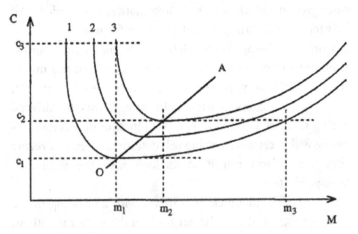

Figure 1

318

racing fitness yielded by various mixes of C- and M-values; a mix with a lower C-value and higher M-value than another on the same isobar will mean a tendency to complete the circuit more quickly balanced by a tendency to incur more penalties. (The numbers attached to the fitness-isobars have only an ordinal significance.) The line OA, which need not have been straight, represents pairings of given C-values each with its optimum M-value. (For a given M-value there is no optimum C-value: an increase of C is never disadvantageous.) Control capacity is running ahead of motor power in combinations above the OA line, and lagging behind in combinations below it.

I call this the Spearhead Model Mk II. Let us now compare it with the Mk I model. To the Mk I idea of a balanced state (i) there corresponds here a position on the OA line. To the Mk I idea of a shift to an over-powered state (ii) there corresponds here a horizontal move to the right away from a position on the OA-line, and to the Mk I idea of a shift to an under-powered state (iii) there corresponds here a vertical move upwards. It is with these unilateral moves away from an optimum combination that important differences open up. In the Mk I model, a unilateral increase in M, even quite a small one, drives fitness right down, while a unilateral increase in C, even quite a large one, leaves fitness unchanged so long as there is no increase in M. In the Mk II model, unilateral increases of M away from a position on the OA-line reduce fitness, but in a continuous, non-abrupt manner, while unilateral increases of C normally bring *actual* as well as potential benefit. Thus with M remaining constant at m_1 an increase of C from c_1 to c_2 immediately raises fitness from level-1 to level-2; and this increase of C is also potentially beneficial in that a subsequent increase of M from m_1 to m_2 with C now remaining constant would raise fitness from level-2 to level-3. The reason why the isobars eventually become vertical on the left is that, when the value of M relative to C is low enough, there comes a point where further increases in C are unavailing. (Reduced to driving a minicab, our Grand Prix champion might perform no better than our minicab driver.) Thus with M stuck at m_1, increases of C beyond C_3 would bring no actual benefit.

That an enhancement of control capacity almost always has actual as well as potential utility is the crucial difference of the Mk II from the Mk I model. If we look for real cases where an increase of control capacity has *only* a potential utility, I think that they will be found to presuppose the possibility of saltations on the motor side. Years ago I had a typewriter with a black/red control lever which was of no actual utility because the ribbon was all-black. More recently I had a computer with a colour control which was of no actual utility because the display unit was black-and-white. To become advantageous these controls called for a saltation to a black-red ribbon, or to a colour display unit. Saltations are of course foreign to evolutionary developments as under-stood by Darwin. The continuities which characterize these should ensure that any enhancement of control capacity that improves it *vis-à-vis* increased motor power in the future will also improve it *vis-à-vis* existing motor power: no potential advantage, we might say, without actual advantage.

Of course, this Mk II Model, with its assumption that developments on the motor side take place along only one dimension, is still terribly over-simplified. It is as if we attended only to our leopardess's fleetness, ignoring her claws, teeth, etc. But my hope is that elaborating it to reflect the multi-dimensionality of real biological motor systems would leave unchanged its central message, which is that *advances in central control systems lead the way in evolutionary developments*. This now receives a straightforward Darwinian endorsement. Unilateral increases in C will nearly always be selected for since they are never unfavourable and nearly always *actually* as well as potentially favourable. By contrast, a unilateral increase in M will be favourable only if the existing combination of C and M is above the OA line. An evolutionary development would be selected against if it strayed into the area below the OA line, but there is nothing to stop it rising way above that line. Control capacity may run far ahead of motor power, but motor power may develop only along paths marked out by its senior partner. And we can say to Wallace that slight variations *could* be of real use: a slight increase in motor power that fits in with pre-existing control capacity is likely to be exploited skilfully and intensively. Start with a number of equally skilled Grand Prix drivers in identical racing machines, and now give one car a very slight improvement to its acceleration; its driver will exploit this to the utmost, and in a close race it might just edge him into first place.

III

As a way of bringing out the philosophical significance of the Spearhead Model I turn now to differences between the views put forward by respectively Karl Popper and John Eccles in their joint work, *The Self and its Brain*. Both men upheld a brand of Cartesian dualism and interactionism; they both maintained that there is a ghost in the machine, and that the ghost significantly influences the machine. (I agree with that.) But Popper held an evolutionist version of Cartesian interactionism, whereas Eccles reverted to something close to the classical dualism of Descartes himself. With Descartes, an immaterial soul is sent into a naturally generated body from without. Eccles said that his body is a product of evolution and natural selection, but his soul or self-conscious mind had a supernatural origin (*SB*, pp. 559–560). My thesis will be that this divine-origin hypothesis, as we may call it, is afflicted by a serious difficulty which is dissolved by the Spearhead Model, at least in its Mk II version. Unfortunately, the fundamental disagreement between Popper and Eccles surfaced near the end of their book (pp. 559f.) and soon fizzled out. There is no mention of the Spearhead Model in *The Self and Its Brain*; it seems that by then Popper had silently abandoned it.

I will now explore the Descartes–Eccles position from within for internal difficulties, engaging only in immanent criticism and not questioning its main assumptions. Descartes had envisaged the soul playing upon the pineal gland whose movements control the flow of animal spirits and thence muscular contractions and bodily movements. He claimed that in human beings the pineal gland, being very small, is easily moved. Eccles, in answer to the question: 'How does my soul come to be in liaison with my brain that has an evolutionary origin?' (*SB*, p. 560), gave to what he called the 'liaison brain' a role quite analogous to that of the pineal gland in Descartes's system. The main difference is that for Descartes the pineal gland has a fixed location within the brain, whereas for Eccles this liaison role is played by different parts of the brain at different times. Eccles claimed that the liaison brain, being at any given time only a minute fraction of the whole brain, is susceptible to the 'weak actions' by means of which the mind intervenes upon the neural machinery to bring about voluntary actions (SB, pp. 356–364).

The great difficulty for dualist interactionism is of course how something immaterial can affect something physical. But suppose that that difficulty can be overcome, perhaps along Humean lines. Then a further problem arises which has not been much noticed.

Descartes said that the *slightest* movement in the pineal gland may alter the course of the animal spirits *very greatly.*[7] So if the soul were to make small errors in the movements it transmits to the pineal gland; these might get amplified disastrously; how does it avoid such errors? We may call this the 'sure touch' problem. It arises equally for Eccles. He wrote: 'Presumably the self-conscious mind does not act on the cortical modules with some bash operation, but rather with a slight deviation. A very gentle deviation up or down is all that is required' (SB, p. 368). It is good that the mind is not a brain-basher but treats the brain gently. But how is the mind able to get its 'very gentle' deviations just right?

---

[7] *Passions*, 34.

A three year-old child asks: *Where those dogs goed?*[8] The grammar isn't quite right but there's nothing wrong with the vocalization. Typically, small children do not commit the vocal equivalent of a typo, hitting the wrong key as it were and inadvertently emitting *Where fose bogs koed?* or whatever. How on the divine-origin hypothesis does our child's soul achieve this nice control over larynx, lips, tongue, etc.? It has been in liaison with its brain for only quite a short time. I now have a nice (well, quite a nice) control over my word-processor, but in the early days I made all sorts of mistakes and spent a lot of time consulting guide-books and pestering friends for advice. But there are no guide-books or advisors to tell the child's soul just which cortical modules to play upon, and just what gentle deviations to give them. Has the divine-origin hypothesis any explanation for this remarkable competence of the soul of our three year-old with respect to its cerebral cortex?

The suggestion that a newly embodied soul proceeds to acquire this competence by trial and error runs up against the enormous complexity of the cerebral cortex. If you were a newcomer to a space-ship's control cabin, and found yourself confronted with 10,000 controls each with 10,000 positions, you would surely abandon all hope of acquiring control of the ship through a trial and error process. By Eccles's estimates, the cerebral cortex consists of something of the order of 10,000 modules each containing something of the order of 10,000 neurones (*SB*, pp. 228, 242). It would

[8] I take the example from Roger Brown.

seem that the only remaining possibility allowed by the divine-origin hypothesis is that when a soul is sent into a body it is already divinely pre-attuned to the control tasks that lie ahead of it. But this would mean that the soul, on entering the body, is not the pure and simple substance traditionally assumed by adherents of the divine-origin hypothesis, but is dependent, at one remove, on antecedent natural processes. If He is to pre-program a soul appropriately before sending it into a body, God will need a detailed plan of the naturally evolved brain to which the soul is to be pre-attuned; for instance, to prepare the soul for control over the speech apparatus He would presumably need, among other things, a map of Broca's area. The pre-attunement hypothesis turns God into a middleman between biological evolution and human mentality.

Armed with the Spearhead Model Mk II, one can say: *Je n'ai pas besoin de cette hypothèse*: there is no need for a Third-Party to prepare the mind for control over the body if they have been evolving together, and moreover the control system has been leading the way with the motor system following in its wake. True, the control-capacity/motor-power dualism is not the same as the mind/body dualism, and the Spearhead Model on its own says nothing about the extent to which a hominid's conscious processes are part of its control system. But the evolutionist theory within which the Spearhead Model was developed provides a cogent argument for 'the efficacy of the mental'. It runs as follows. Major premise: *if* a phenotypic character appears on the evolutionary scene relatively late and subsequently spreads widely among a great variety of species, becoming more strongly pronounced during phylogenetic development, *then* there is an overwhelming presumption that it is being selected for its survival value. Minor premise: consciousness satisfies the antecedent clause in the above premise. Lemma: something can have survival value for an animal only if it affects its bodily performance. Conclusion: there is an overwhelming presumption that consciousness affects bodily performance.

True, a mental event's having efficacy does not guarantee it a place in the control system. Suppose that you feel a sudden stab of pain for which there is no obvious explanation; then it goes away without seriously disturbing what you were doing, and you forget about it. A month later this happens again, and you again forget about it. But when it happens a third time you decide to see your doctor. In this case we might say that the pain had a certain efficacy—it helped to get you to the doctor's waiting-room. But it did so by nudging your control system from without, as it were. But now consider a champion billiard-player preparing to make a stroke. After surveying the table to select the most promising combination, he sights along his cue, mentally estimating the required collision angle and momentum, positions his body just right, and makes a nicely controlled shot—and the balls move just as he planned. He is exploiting skills developed by our hominid ancestors playing survival games rather billiards. In cases like this, thought processes surely are playing a crucial role within the control system whatever materialists and epiphenomenalists may say. Popper was surely right when he said that 'the biological function of the mind is clearly closely related to the mechanisms of control' (*SB*, p. 114).

I introduced the Spearhead Model with the example of a leopardess procuring a meal for her cubs. She too behaved as if she could estimate the (in her case rapidly changing) situation accurately by eye, make lightning calculations from relevant formulas, and then get her motor system to execute the result immaculately. But here a big difference opens up between animal and human control. Any other healthy leopardess of similar maturity would probably have performed about as well, whereas very few other human beings could perform anything like as well as our champion billiard player. Descartes remarked that, while many animals exhibit in some of their actions a specialized dexterity which we cannot match, we exhibit a versatility and, he might have added, the ability to foster and develop dexterities, which they cannot match (*Discourse*, p. 117). We might put alongside the dexterity of our champion billiard player the dexterity exhibited by a concert pianist playing Chopin, say. It cannot be seriously doubted that these people possess a degree of control over their limbs, especially their fingers, that is supererogatory from the standpoint of biological survival and reproduction. This does not create a new difficulty for the divine-origin hypothesis. (Some difficulty may be created for it by the fact that ordinary folk lack this endowment: should not the control exercised by divinely manufactured souls be *uniformly* excellent?) But it might seem to create a difficulty for a naturalistic world-view which gives a central place to Darwinian evolution. Houdini said: 'I have to work with great delicacy and light-ning speed', adding that he had had 'to make my fingers super-fingers in dexterity, and to train my toes to do the work of fingers'.[9] How could natural selection have endowed him with the ability get his toes working skilfully for him?

I think that the Spearhead Model, when coupled with the fact that there has been an astonishing threefold increase in the size of hominid brains over the last three million years, enables us to handle this difficulty very satisfactorily. The most widely held explanation for the rapid swelling of hominid brains invokes sexual selection, and I will go along with that. With sexual selection it can happen that a runaway development starts with members of one sex sensibly going for a biologically desirable feature x in the opposite sex but which is then so persistently enhanced that a time comes when further enhancement of x has no biological utility. The Spearhead Model shows how this could happen if x is, or is correlated with, control capacity. For it shows that, with motor power M remaining constant, persistent increments of control capacity C will go on having biological utility until a level is reached above which enhancements of C have no biological utility. Natural selection could not drive C above that level, but sexual selection might; and if it does, genes for various kinds of exceptional control capacity may have got into the human gene-pool without necessarily sweeping through it and without being driven out, and getting picked up occasionally.

[9] Harold Kellock, *Houdini: His Life Story* (Heinemann, 1928), p. 3.

11 Popper and the Scepticism of Evolutionary Epistemology, or, What Were Human Beings Made For?

MICHAEL SMITHURST

There is a sort of scepticism, or, at least, epistemological pessimism, that is generated by appealing to Darwin's theory of evolution. The argument is that nature, that is the selective pressures of evolution, has clearly fitted us for certain sorts of learning and mundane understanding, directly beneficial in point of individual survival and chances for reproduction. Very likely then, it is argued, nature has not fitted us for arcane intellectual accomplishments remote from, or quite disconnected from, those ends. So, it is suggested, perhaps we cannot understand, perhaps never will understand, because we are not made to understand, such matters as consciousness, its nature and causes, the origins of life, the beginning of the universe, or astro-physics in its more finalist pretensions. Sometimes, taken with the claim that manifestly we do understand some of these things, the argument becomes a *reductio*, and its exponents claim that consequently Darwinism must be false, or of only limited application. More commonly, however, in 'naturalized epistemology', it is made the vehicle of a claim about the limits of science, limits imposed by the evolutionarily derived frailty of human understanding. Popper, the late Popper, is one of the progenitors of naturalized epistemology. I want to ask what Popper's response would be, or ought to be, to this epistemological nihilism (to purloin a phrase of Quine's) conjured out of Darwinism.

The particular twist that Popper gave to evolutionary epistemology took the form of 'thesis Darwinism',[1] the proposal that an analogue of natural selection, in the form of critical attempts at falsification, works on competing theories to allow only the fittest to survive. By contrast, the main philosophical use of evolutionary theory in recent years has been with respect to the character and development of human cognitive capacities. Darwinism fills a long felt need there. Hume, having expounded his sceptical doubts concerning the understanding, to the effect that all inductive reasonings beg the question by presupposing (what J. S. Mill later denominated) the uniformity of nature, and having offered his own 'sceptical solution' to those doubts, namely, that it is our nature to reason by similitudes from past to future, and that custom or habit, not reason, is the foundation of so doing, observed:

> Here then is a kind of pre-established harmony between
> the course of nature and the succession of our ideas; and
> though the power and forces by which the former is
> governed be wholly unknown to us, yet our thoughts and
> conceptions have still, we find, gone on in the same train

---

[1] The expression is Nicholas Rescher's. See N. Rescher, *A Useful Inheritance: Evolutionary Aspects of the Theory of Knowledge* (Savage, Maryland: Rowman & Littlefield Publishers, 1990), ch. 2.

> with the other works of nature... Those who delight in
> the discovery and contemplation of *final causes* have here
> ample subject to employ their wonder and admiration.[2]

Hume did not have the resource of Darwinism available to him. Quine and others have since applied it. In spite of Popper, it has become common to say that man is an inductive animal, and to argue that induction is an inherited cognitive propensity which has persisted and proliferated because it runs with the grain of nature and gives us 'a lien on the future'.[3]

Popper's use of Darwin is quite different. He is struck by the Darwinian figure of life proceeding in its developments and adaptations by trial and error. This parallels the method of conjecture and refutation that he had already identified as the logic of scientific discovery. So Popper makes theories, not cognizing individuals, the subjects of the evolutionary process:

> the growth of our knowledge is the result of a process
> closely resembling what Darwin called 'natural selection';
> that is, the natural selection of hypotheses; our
> knowledge consists, at every moment, of those
> hypotheses which have shown their (comparative) fitness
> by surviving so far in their struggle for existence; a
> competitive struggle which eliminates those hypotheses
> which are unfit.

[2] D. Hume, *Enquiries Concerning Human Understanding and Concerning the Principles of Morals* (Oxford: Clarendon Press, 1975), pp. 54–55.

[3] W. V. Quine, 'Natural Kinds', in W. V. Quine, *Ontological Relativity and Other Essays* (New York: Columbia University Press, 1969).

This interpretation may be applied to animal knowledge, prescientific knowledge, and to scientific knowledge. What is peculiar to scientific knowledge is this: that the struggle for existence is made harder by the conscious and systematic criticism of our theories. Thus, while animal knowledge and pre-scientific knowledge grow mainly through the elimination of those holding the unfit hypotheses, scientific criticism often makes our theories perish in our stead, eliminating our mistaken beliefs before such beliefs lead to our own elimination.

This statement of the situation is meant to describe how knowledge really grows. It is not meant metaphorically, though of course it makes use of metaphors. The theory of knowledge which I wish to propose is a largely Darwinian theory of the growth of knowledge. (*OK*, p. 261)

It is hard to take seriously Popper's claim that what he says here is not meant metaphorically. The ontological literalism of his 'World 3' mode of discoursing about theories is perhaps what prompts him to say it. Thesis Darwinism is nonetheless metaphorical, and the metaphor is shot full of holes.

Except in the most figurative sense, theories do not reproduce, and hence they do not reproduce with Malthusian fecundity. A key to speciation is reproductive isolation, with its implied avoidance of conflict, conflict that the critical method for theories must always invite. Popper sees theory-elimination as a progressive enterprise, for it is a matter of getting nearer to the truth; but Darwinian evolution is non-progressive and getting nearer to nothing, a point that, ironically, T. S. Kuhn, who also represents his philosophy of science as 'evolutionary', invokes as a reason for eschewing mention of truth in talking about theories.[4]

Empedocles is credited with a theory of evolution which posited that the gods created a profusion of kinds, cows with human heads, men with eyes where ears should be, and ears in place of arms. What survived is what from this initial variety was by chance fitted to survive. Such a theory of evolution is a 'hopeful monsters' theory. It is natural selection without Darwinian gradualism. Science is full of hopeful monsters, but nature is not.[5]

As is evident in the above quotation from *Objective Knowledge*, Popper does sometimes run thesis Darwinism together with evolutionary assumptions about the development of human cognitive capacities. In fact, Popper puts an evolutionary slant on his reading of Kant. He says in *Realism and the Aim of Science* (1983) that the 'Kantian psycho-physiological digestive mechanism with which we are endowed' causes certain facts, facts about causality and identity, facts which really arise from our own psychic or physiological apparatus, to be 'imposed by us upon the world, in the sense that they are bound to become

---

[4] T. S. Kuhn *The Structure of Scientific Revolutions* (second edition, Chicago: University of Chicago Press, 1970), pp. 170–173.

[5] For a good discussion of Popper's 'thesis Darwinism' see A. O'Hear, 'On What Makes an Epistemology Evolutionary', *Proceedings of the Aristotelian Society*, Supplementary Volume LVIII (1984), 193–217. For the comparison of theories to hopeful monsters, see John Watkins in this volume.

"objective" laws of the things which we perceive'. Popper then continues:

> And it might be conjectured that our belief in real things is similarly physiologically founded. But in this case, the physiological mechanisms and the beliefs which spring from them (both, are, we may conjecture, the results of a long evolution and adaptation), seem to withstand criticism and to win in competition with alternative theories. And when they do mislead us, as in a cinema—especially in cartoons—they do not lead adults to assert seriously that we have before us a world of things. Thus we are not (as Kant and also Hume thought) the victims of our 'human nature', of our mental digestive apparatus, of our psychology or physiology. We are not forever the prisoners of our minds. We can learn to criticize ourselves and so to transcend ourselves. (RAS, p. 154)

This quotation, I claim, and similar ones which can be adduced, put Popper into the, as it were, optimist camp, against the naturalistic sceptics. Popper also argues that the values and beliefs of a social group may be a similar bondage, and 'may also have a physiological basis'. But these barriers of culture and biology, in his opinion, are surmountable: 'Rational discussion and critical thinking are not like the more primitive systems of interpreting the world; they are not a framework to which we are bound and tied. On the contrary, they are the means of breaking out of the prisons—of liberating ourselves'. (RAS, p. 155)

I want to interject at this point a reminder of Popper's fallibilism, and the reasons behind it. Fallibilism hovers between the modest thought that any theory might be wrong, and the contention that every theory will at some point be shown to be in error. Why does Popper believe in scientific fallibilism? Many hold to it because of the fallibilist induction. This surveys the history of science, and infers: one thing is certain, every theory has eventually been overthrown; the fate of every scientific theory is ultimately to be replaced by a better theory. Put like that, the fallibilist induction is not an attractive argument. First, the induction chases its own tail, theories endlessly succumbing to new theories. Secondly, the history of science as a self-consciously distinctive enterprise has been only about four hundred years, and so the inductive base is not large. Thirdly, depending on what counts as 'theory' and how theory sits with 'discovery', it is by no means obvious that all scientific theories have been overthrown, or eventually will be. Harvey on the circulation of the blood, for instance? And presumably those physicists dreaming of a Final Theory do not accept the conclusion. Popper himself, of course, makes no inductions, and so he does not make the fallibilist induction.

There are two sources of Popper's own fallibilism. One is his conception of scientific method, according to which theories, their status being that of permanent conjectures, are not properly objects of belief at all, hence not objects of infallible belief. Another is the particular way in which he conceives of science as always open-ended. Popper alleges that, in advanced science, a newly successful theory is one that by its very nature (it is conceptually innovatory) always raises new problems. A successful new theory, reconceptualizing and embracing the predictive successes of the theory it supersedes, is an advance into new and unantici-patable complexity. New problems can only be cracked by conceptual imagination. We cannot create an armoury in advance for future use, 'problems can only be solved with the help of new ideas'. In *Unended Quest* he gives this advice:

> do not try in advance to make your concepts or formulations more precise in the fond hope that this will provide you with an arsenal for future use in tackling problems that have not yet arisen. They may never arise; the evolution of the theory may bypass all your efforts. The intellectual weapons which will be needed at a later date may be very different from those which anyone has in store. (*UQ*, p. 30).

Sniping at the analytical philosophy of his contemporaries and their pretension that the proper function of philosophy is conceptual analysis to assist the progress of science, he points out that no one would have hit on Einstein's analysis of simultaneity in advance of Einstein's problems. Popper describes science as 'revolution in permanence' and has no sympathy for Kuhn's notion of 'normal science'. Exercises of creative imagination are central to Popper's understanding of theoretical endeavour. I mention the point in order to revert to it later.

I turn now to some statements of the sceptical naturalist apprehensions. Michael Ruse in *Taking Darwin Seriously* provides some good examples. For instance, 'the nature and development of science is constrained and informed by the biologically channelled modes of thinking imposed on us by evolution'. Ruse conjectures that logic and mathematics are grounded in innate principles, principles that have spread because they confer survival advantage: 'Those humans who believed that 2+2=5, or that fire causes orgasms rather than pain, or ignored the virtues of inductive consiliences, got wiped out in the struggle for existence.'[6] Ruse points out that 'natural selection simply does not care about giving us a meticulously true and comprehensive insight into the nature of things'. He even argues that the 'necessity' that most philosophers have found in logic and mathematics may be just a natural device for reinforcing logical and mathematical beliefs: 'biological fitness is a function of reproductive advantage rather than of philosophical insight. Thus, if we benefit biologically by being deluded about the true nature of formal thought, then so be it. A tendency to objectify is the price of reproductive success'.[7]

Hume was the first philosopher to argue that metaphysical beliefs in general, and sceptical ones in particular, rarely or never make any difference to anything. They are disconnected from action, and the ideas embodying them can only momentarily be infixed with the liveliness that Hume equates with belief. Ruse agrees, and has his own explanation of this: 'We are animals, and have adaptations to protect us from the worries of reason. If we get overly depressed about the conclusions of our thinking, we would simply stop functioning properly.'[8]

[6] M. Ruse, *Taking Darwin Seriously: A Naturalistic Approach to Philosophy* (Oxford: Blackwell, 1986), p. 189.
[7] Ibid. p. 172. [8] Ibid. p. 188.

Ruse repeats a common conjecture that 'primitive peoples' have certain bodies of false belief, their religions for example, 'because such beliefs have adaptive value'. Similarly, he speculates, our scientific theories might be 'illusions fostered upon us for reproductive purposes'. In fact,

> the Darwinian approach—with its fundamental commitment to non-progressivism—has to lay itself open to radically different interpretations of the world, where nothing we hold as intuitively true or obvious holds. . . At the extreme the Darwinian approach has to admit that everything that we believe may simply be false, even unto the very principles of Darwinian evolutionary theory.[9]

No doubt this is not the most convincing way of making the sceptical naturalist case, for it throws all natural knowledge into question and is confessedly self-undermining. The more inviting way is to take Darwinism for granted, to assume the reliability of some of our evolved cognitive mechanisms, and to insinuate their fragility when we come to address theoretical questions of a practically remote character. The charge becomes that, epistemologically, we are over-reachers.

This is Thomas Nagel's variant of the move in *The View From Nowhere*. In fact, Nagel does not argue for a sceptical conclusion, but reasons, as Alfred Russel Wallace did, that the extent of human intellectual capacity constitutes a case against Darwinism, or, at least, a serious limitation of it: 'if, *per impossibile*, we came to believe that our

[9] Ibid. p. 201.

capacity for objective theory were the product of natural selection, that would warrant serious scepticism about its results beyond a very limited and familiar range'.[10]

The over-reacher argument is used directly to a sceptical end by Colin McGinn. He argues in an article 'Can We Solve the Mind–Body Problem?',[11] that minds are subject to 'cognitive closure'. The mind of a rat and the mind of a monkey have different powers, biases, and blindspots. Consciousness must be a natural property arising from certain organizations of matter, but 'our concepts of consciousness just are inherently constrained by our own form of consciousness'. McGinn brings out that it is not so much the capacity of mind that it is at issue in the sceptical argument from Darwinism, but the character of mind:

> Conscious states seem biologically quite primitive, comparatively speaking... It is not the size of the problem but its type that makes the mind–body problem so hard for us. This reflection should make us receptive to the idea that it is something about the tracks of our thought that prevents us from achieving a science that relates consciousness to its physical basis.[12]

McGinn's argument involves more than merely an appeal to Darwinism, but he shares Ruse's fear that, in some respects

---

[10] T. Nagel, *The View From Nowhere* (Oxford University Press, 1986), p. 79.

[11] C. McGinn, 'Can We Solve the Mind–Body Problem?', *Mind*, **xcvii**, no. 981, (1989), reprinted in C. McGinn, *The Problem of Consciousness: Essays Towards a Resolution* (Oxford: Blackwell, 1991).

[12] Ibid. p. 19.

at least, scientific enquiry involves 'probing into mysteries for which our biology has not fitted us—nor is there reason to think that it ever could fit us'. Though recently revived, the argument is not new. Peirce put it sparingly as: 'on unpractical subjects natural selection might occasion a fallacious tendency of thought'.[13]

How might the scepticisms of evolutionary epistemology be resisted? Some versions misconceive Darwinian evolutionary theory. I have criticized elsewhere Nagel's variant of the ploy.[14] It is objectionable for two reasons. First, it is not true that a Darwinian is committed to explain human intelligence as adapted to a range of concrete immediate survival skills. That is indeed the surface at which natural selection must always operate, and for that reason the inference is tempting, but there are plausible alternative hypotheses. Secondly, Nagel's version of the argument falls into the fallacy of 'hyperselectionism'. It exaggerates adaptedness and assumes that living creatures are finely honed by natural selection, brought over time by a multitude of selective pressures to being near ideally made for those functions that enable them to survive and reproduce. This was Wallace's view, but it was not Darwin's. Living forms are more ramshackle constructions than that. Evolutionary success holds only from moment to moment. A species survives only in

---

[13] C. S. Peirce, *Collected Papers*, Vol. V (Cambridge, Mass.: Harvard University Press, 1934), sec. 5.366. Quoted by Rescher in *A Useful Inheritance*.

[14] M. Smithurst, 'The Elusiveness of Human Nature', *Inquiry*, **33**, No. 4 (1990), pp. 1–13.

that the individuals comprising it do so. An individual's mode of survival may be circuitous and uneconomical, but that is inconsequential provided survival and reproduction are indeed secured. The recurrent laryngeal nerve in mammals is an illustrative example. This nerve runs from brain to larynx, not directly, but by passing down the neck to the chest, then returning up again. Its vulnerable and inefficient length results from the fact that, in aquatic ancestral forms, the nerve had to take a path around the gill arches. Those once restricting anatomical features are now highly modified, but in the modern camel the nerve still twice traverses the length of the animal's long curved neck. Other adaptations would be better, but they could only be achieved by an embryologic reorganisation of the species. Once a nexus of characteristics is in place, innumerable constraints follow about what developments are then possible. Pigs can't ever have wings.

Not all defenders of naturalistic scepticism infer that Darwinism entails that the human mind must be prosaically limited. Nicholas Rescher, for example, argues that over-capacity of intellect is needed both to deal with ordinary survival problems quickly, and to deal with the really difficult problems when they do occur. This leaves us with an over-capacity 'to pursue various challenging projects that have nothing whatever to with survival'.[15] Rescher also addresses the question why we humans are not more intelligent than we are: 'As the numbers of "clever people" who

[15] Rescher, *A Useful Inheritance*, p. 110.

pride themselves on strength of intellect increases, social cohesion becomes more difficult to obtain. University faculties are notoriously difficult to manage.'[16] True, no doubt, but neither societies, nor even university faculties, are units of natural selection. Haplodiploidy, linking us to a common mother, is all that can make a 'society' a unit of selection.

Pace Michael Ruse, I do not believe that there ever has been any group of proto-humans who believed that 2+2=5, and suffered extinction as a result. '2+2=5' and '2+2=4' cannot stand alone. They are possible beliefs, if that is what they are, only within the context of a system of arithmetical principles and practices. There is no sense in the supposition that a creature might have just one arithmetical belief, and that a false one. 'Neural Darwinists' have proposed all manner of surprising modules as relatively self-contained, genetically based, constituents of mind. The example alleged of a British family with an hereditary inability to form plurals, replicates itself through the linguistics textbooks as ubiquitously as the Indian prince, refusing to credit that water turned into ice, wandered through seventeenth- and eighteenth-century philosophy. But even on such a theory, '2+2=5', 2+2=4', either will not modularize. The granular physical independence of the genotype requires a corresponding logical independence of the phenotype, and, with arithmetic, you cannot have it.

Rescher speculates about alien life forms with their own systems of science. He says that, because scientific

---

[16] Ibid. p. 116.

information is an 'ideational construct', the sameness of the object does not guarantee the sameness of the thoughts about it. Such beings would deal with the same world as us, but might 'differ in mode of formulation, in subject-matter orientation, and in conceptualisation'. Natural science, says Rescher, 'does not depict "reality as such" but rather affords us a picture of "reality as it presents itself to us"—we being inquirers of a certain particular sort, with a certain particular evolution-determined mode of emplacement in the world's scheme of things'.[17] So it follows that for rational alien beings 'the taxonomic and explanatory mechanisms by means of which their cognitive business is transacted could differ so radically from ours that intellectual contact with them would be difficult or impossible'.[18]

This is, of course, an old acquaintance, the impenetrable, incommensurable conceptual scheme. Donald Davidson's argument against the very idea of a conceptual scheme should be recalled. Davidson argues that putative differences of conceptual scheme amount to differences of language, and nothing recognisable as a language is in principle untranslatable. Translation requires the attribution of beliefs, desires, and intentions. To make such attributions we must hypothesise that, in the main, the putative speaker believes what we believe:

> Nothing... could count as evidence that some form of activity could not be interpreted in our language that was

[17] Ibid. p. 77.   [18] Ibid. p. 85.

not at the same time evidence that that form of activity was not speech behaviour.[19]

What makes interpretation possible, then, is the fact that we can dismiss *a priori* the chance of massive error. A theory of interpretation cannot be correct that makes a man assent to very many false sentences; it must generally be the case that a sentence is true when a speaker holds it to be.[20]

Successful communication proves the existence of a shared, and largely true view of the world.[21]

Let me put Davidson's point in a way that is not at all his, but illustrates his central contention in a way that suits my purposes. Consider an alien being of physiologically different aspect from ourselves, multiple eyes, prehensile tongue, tentacles instead of arms, and the rest of it. We should not be in the least surprised to find that this creature differed from us in its perception of the so-called secondary qualities. Indeed, that such a being should be sensitive to parts of the spectrum that we are not, that it should hear sounds at pitches inaudible to us, and that it should taste bitter where we taste sour, savour what disgusts us and repudiate what we relish, is just what we would expect. The basis of our species-peculiar subjectivities could be explained by physiology and

---

[19] D. Davidson, 'On the Very Idea of a Conceptual Scheme' in D. Davidson, *Inquiries into Truth and Interpretation* (Oxford: Clarendon Press, 1984), p. 185.

[20] D. Davidson, 'Thought and Talk', in S. Guttenplan (ed.), *Mind and Language* (Oxford University Press, 1975), p. 21.

[21] D. Davidson, 'The Method of Truth in Metaphysics', in Davidson, *Inquiries into Truth and Interpretation*, p. 201.

psychology, and it is quite conceivable that we could agree with the aliens about our differences and agree on the explanations of them. But consider now the suggestion that what hops or rolls down the ramp of some inter-stellar craft should systematically differ from us in respect of judgments about the so called primary qualities—shape, size, position, and number; differ from us in judgments, that is, as manifested in the creature's behaviour, in its address to objects. We simply have no idea of what we are invited to imagine. Suppose I judge that there are two billiard balls here on the table, that they are stationary, round, of course, and about the size that billiard balls standardly are. This shows itself in where I look, how I pick them up, and in the fact that I do not come back for more when the two are taken. But what behaviour am I now to imagine on the part of the alien to suppose it judging, in respect of these same objects, that they are not at rest but in motion, not on the table but against the ceiling, not two in number but five, square not round, and each as big as a hatbox? No 'behaviour' is imaginable, only movements that strike the air. Agreement in judgments about the primary qualities is critical to recognizing another creature as sensitive, let alone rational. This is the germ of truth in the old distinction between primary and secondary qualities, despite the rottenness of the classical arguments.

Rescher amusingly supposes scientific moles in a subterranean civilisation, considers the perennial favourite, rational dolphins, and speculates on intelligent inter-galactic clouds. Rational moles, porpoises, and science as it might be in the Kingdom of the Blind, are all fairly easily dealt with. There are no logically insuperable barriers to communication there. But what of rational inter-galactic clouds? A story might be told that makes them analogous to the above cases, but if not, one might as well attach the word 'rational' to a teapot, and say teapots might be rational. —'We only say of a human being or what is like one that it thinks', said Wittgenstein. 'Only of a living human being and what resembles (behaves like) a living human being can one say: it has sensations; it sees; is blind; is deaf; is conscious or unconscious.'[22]

Does it in fact make sense to lament that evolution might not have fitted us to understand consciousness, astrophysics, or whatever?

There are problems which are set by the ambient environment; such as, how to stay afloat, how to get at those grubs which are good to eat, how to avoid that approaching raptor. There are also problems which we propose. Scientific problems are projective concerns. They do not exist in the world as problem features of the world.

It makes sense to say that we, human beings, are not equipped by evolution to break open walnuts, but those other creatures are. It makes no sense to say that they are equipped by evolution to understand astro-physics, or the origin, of life, or consciousness, but we, alas, are not. It makes no sense, because evolutionary theory, as it stands, gives us no conception of what it would be like for evolution to mould creatures to such ends. The notion of natural selection does not catch here. Having a good astro-physical

[22] L. Wittgenstein, *Philosophical Investigations* (Oxford: Blackwell, 1963), paragraphs 360 and 281.

theory might or might not have some remote connection with survival. Who knows? But the problems of survival that are set by nature, and which drive the evolution of species, are immediate problems. These are the only problems that evolution deals in. Such is the very nature of the theory.

Unlike, are we made to crack walnuts?, the question, are we made to understand astro-physics?, is misconceived. Yet the naturalistic sceptic sees no difference between these questions. He makes their equation a springboard for his scepticism: 'You were made to crack walnuts, etc., so whatever gave you the hubristic idea that you might be made to understand astro-physics or explain consciousness?' But, from the evolutionary point of view, there is not anything we cannot do here, a something that, perchance, evolution has not fitted us for. Maybe we can crack the problem of consciousness, maybe we cannot. In that sense it may be something we cannot do, but evolution has nothing to do with it, one way or the other. Complaining that evolution might not have fitted you to understand consciousness is a bit like complaining that a cookery book might prove not to contain instructions on how to construct a television set.

What would it be like to be able to say of some other kind of creature that it understands consciousness, the origins of life, and so on, but we cannot? I know when to say of some bird or insect that it perceives what we cannot perceive, the ultra-violet end of the spectrum, say, and the colours therein. There is nothing problematic about that. But under what circumstances is one entitled to say of other rational beings that they understand what we logically cannot? Well—they might have a more impressive technology than our own, but that does not go far in answering the question. The connection of technology to theory is shifting and remote. These beings of evolved understanding might be less like the technologically invested autarchs of an Evil Empire, and more like the sages of Bernard Shaw's *Back to Methuselah*, old, anorexic, and uncommunicative, and supposedly captivated only by their own rich inner lives. What other picture does anyone have of the bogeyman Superior Understanding?

What does contemporary evolutionary theory have to say about the emergence of human intelligence? Precious little on mundane survival skills, but a good deal, by implication, about creativity and imagination.

The mechanism of big brain development is almost certainly neotony, that is, the holding on to infantile forms. Since Bolk first identified it in the 1920s, under some ridicule, paedomorphic retention has been recognized as a real, if curious, strategy in evolutionary change, one that can conjure large developments from minimal genetic alteration.[23] Premature sexual maturity of the juvenile form of an animal results in offspring who retain in adulthood some early developmental stages. Hence our big embryo heads disproportionate to our bodies. Physiologically, the basis of our intelligence arises from infantilisation. The assumption of an accompanying infantilism in aspects of behaviour, led Konrad Lorenz, who first conceived the notion, and Stephen

---

[23] The history of the idea is detailed in S. J. Gould, *Ontogeny and Phylogeny* (Cambridge, Mass.: Harvard University Press, 1977).

Jay Gould, who has made it generally familiar,[24] to see our experimentalism as that of play, and suggest an image of man as a permanently ludic ape.

Neotony is the mechanism, but what is the reason for big brain development? What fans it? Darwin's own answer is back in fashion, the peacock's tail, or sexual selection. Cumbersome decorative ornaments begin their evolutionary histories as minor and accidental sexual attractants, a tail slightly larger than others, a colour slightly more striking. Consequent reproductive success multiplies the relevant genes, the gene for the accidental characteristic, and the gene for the accidental female preference that made it an attractant in the first place. As, through the chances of genetics, males 'compete' with one another to produce ever more striking specimens of the attractant characteristic, so, over the generations, the characteristic, and the corresponding appetite for it, magnify. It is suggested that intelligence, through the diverting and imaginative performances that it enables, is just such an attractant, a peacock's tail.

Geoffrey Miller criticizes conventional theories of the emergence of intelligence. In respect of most of the ends proposed for rationality, Miller thinks, like Kant, that 'instinctual arrangements' could usually do the job better. He argues that

> the neocortex is not primarily or exclusively a device for toolmaking, bipedal walking, fire-using, warfare, hunting,

---

[24] S. J. Gould, 'The Child as Man's Real Father', in S. J. Gould, *Ever Since Darwin* (Harmondsworth: Penguin Books, 1980).

gathering, or avoiding savanna predators. None of these postulated functions alone can explain its explosive development in our lineage and not in other closely related species... The neocortex is largely a courtship device to attract and retain sexual mates: its specific evolutionary function is to stimulate and entertain other people, and to assess the stimulation attempts of others.[25]

There is more to the story. Sexual selection puts us in a tournament of the mind with others. Since Richard Alexander proposed that 'only human beings themselves could provide the necessary challenge to explain their own evolution',[26] a literature has grown on the theme of humans as 'natural psychologists', adapted as much to the purveying and discovery of disinformation as of disinterested truth. Byrne and Whiten, 1988, in *Machiavellian Intelligence: Social Expertise and the Evolution of Intellect in Monkeys, Apes and Humans*, argue that deception and the detection of deception are primary advantaging characteristics in the encephalization arms race. There is not space here to detail the enigmatic psychological

---

[25] G. F. Miller, 1992. 'Sexual Selection for Protean Expressiveness: A New Model of Hominid Encephalization', paper delivered to the fourth annual meeting of the Human Behaviour and Evolution Society, Albuquerque, New Mexico, July 22–26 1992, cited in M. Ridley, *The Red Queen, Sex and the Evolution of Human Nature* (London: Penguin Books, 1993).

[26] R. D. Alexander, 'The Evolution of Social Behaviour', *Annual Review of Ecology and Systematics* 5 (1974), pp. 325–83. Ridley, ibid. gives a lively survey of the literature, pp. 318–326.

experiments purporting to show that logic is pushed into second place by our permanent will to manipulate information, and its recipients.[27]

We are now in a position to answer the question what were human beings made for? The answer is, to tell lies and to have sex. Is that a cynical and frivolous answer to the question? Certainly not. It is an answer from orthodox Darwinism, under its current speculations. Indeed, considered in a certain light, the conclusion is more edifying than that of the sceptical naturalists. I appeal to Popper to explain why. Popper believes that science proceeds by conjecture and refutation, and conjecture is a matter of the freewheeling use of untrammelled imagination in a non-deterministic world. Indeed, one of Popper's frequently repeated arguments against determinism is that it is crucially at odds with the very possibility of creativity. Under a deterministic system, Popper urges, there could be no true novelty in the world. Originality, he argues, disappears along with freedom in a Laplacean universe, where all invention could, in principle, be predicted in advance. The vehemence of Popper's rhetoric is manifest:

> [Laplacean predestination] leads to the view that billions of years ago, the elementary particles of World

---

[27] L. M. Cosmides, 'The Logic of Social Exchange: Has Natural Selection Shaped how Humans Reason? Studies with the Wason Selection Task', *Cognition*, **31** (1989), pp. 187–276. L. M. Cosmides, and J. Tooby, 'Cognitive Adaptations for Social Exchange', in J. K. Barkow, L. Cosmides and J. Tooby, *The Adapted Mind* (New York: Oxford University Press, 1992). Both cited in Ridley, ibid.

I contained the poetry of Homer, the philosophy of Plato, and the symphonies of Beethoven, as a seed contains a plant; that human history is predestined, and with it all acts of human creativity. And the quantum theoretical version of the view is just as bad. If it has any bearing on human creativity, then it makes human creativity a matter of sheer chance. No doubt there is an element of chance in it. Yet the theory that the creation of works of art or music can, in the last instance, be explained in terms of chemistry or physics seems to me absurd.

(*OU*, pp. 127–128)

Popper's argument is vague and suggestive, and not easy to evaluate. In so far as it engages with the Laplacean demon, the battle has already been won for other reasons. Contemporary 'chaos theory' now allows it to be affirmed *a priori*, using strictly the principles and equations of Newtonian mechanics, that the predictions of the Laplacean demon are impossible, and no amount of inputted information on initial conditions could render them possible. The data on initial conditions must be increased exponentially to edge the predictability horizon forward in linear steps.[28]

To at least the following extent Popper is onto an important point here. Intelligence is arguably a faculty, arguably a unitary capacity, to some degree inherited, and

---

[28] See J. Lighthill, 1986. 'The Recently Recognized Failure of Predictability in Newtonian Dynamics', in J. Mason, P. Mathias and J. H. Westcott (eds), *Predictability in Science and Society* (London: The Royal Society and the British Academy, 1986).

352

with a metric assignable to it. Creativity, I think, cannot plausibly be so represented. On Popper's view, conceptual and explanatory innovation is realized in confrontation with problems, unforseen and unforeseeable problems. Creative imagination is not a resting potentiality, like the shatterability of glass or the sparking propensity of a match.

This claim becomes implausible if it can be shown, perhaps on *a priori* grounds, that human thought must inevitably take certain predictable forms. Carrying the sceptical naturalist argument to a logical conclusion, McGinn has recently produced the conceit of a meta-philosophical meditation on the impossibility of philosophy. He conjectures that philosophy itself is an impossible pursuit because of our intellectual unfittedness for it: 'in trying to do philosophy, we run up against the limits of our understanding in some deep way. Ignorance seems the natural condition of philosophical endeavour.'[29] McGinn asks 'where does the felt profundity of philosophical questions come from?' and 'what specific characteristics of conscious reason put philosophy beyond its scope?'. Following up a speculation of Chomsky that 'our arithmetic faculty is a product of our linguistic faculty, got by abstracting from one domain to another', McGinn adds that, when joined with our modes of spatial representation, particularly the visual, these same hypothesised innate principles could also account for our ability to do physics. Theories we can understand, he

[29] C. McGinn, 1994. 'The Problem of Philosophy', *Philosophical Studies*, 76 (1994), p. 133.

suggests, may be limited to those exhibiting 'combinatorial atomism with lawlike mappings'.

Popper's philosophy of science is characterized not only by a resistance to determinism but by a rooted objection to reductionism. If McGinn's generalized sketch of the form necessary for any invention or theorizing to be accessible to human understanding strikes one as true, Popper's 'creativity' argument will not seem convincing. I will not attempt to evaluate Popper's argument further, but put it in evidence as to where he is to be located on the question raised by the sceptical naturalists.

Finally, I will lend Popper an ally, in the person of William Blake. The aspirant Darwinians of the Philosophy Departments are not the first philosophers to set *a priori* the limits of the human mind. Some did it with a theory of meaning, some with a theory of ideas.

Hume died in 1776. Blake stood at the end of the eighteenth century, on the far edge of the theory of ideas. Hume writes:

> Nothing at first view, may seem more unbounded than the thought of man, which not only escapes all human power and authority, but is not even restrained within the limits of nature and reality. To form monsters and join incongruous shapes and appearances costs the imagination no more trouble than to conceive the most natural and familiar objects... But though our thought seems to possess this unbounded liberty, we shall find upon a nearer examination that it is really confined within very narrow limits, and that all this creative power of the mind amounts to no more than the faculty

of compounding, transposing, augmenting, or diminishing the materials afforded us by the senses and experience.[30]

Combinatorial atomism in fact. Blake, in 1788, wrote this short piece. It is called 'There is No Natural Religion':[31]

The Argument

Man has no notion of moral fitness but from Education. Naturally he is only a natural organ subject to Sense.

I

Man cannot naturally Perceive but through his natural or bodily organs.

II

Man by his reasoning power can only compare and judge of what he has already perceiv'd.

III

From a perception of only three senses or three elements none could deduce a fourth or fifth.

IV

None could have other than natural or organic thoughts if he had none but organic perceptions.

V

Man's desires are limited by his perceptions: none can desire what he has not perceiv'd.

---

[30] D. Hume, *Enquiries Concerning Human Understanding and Concerning the Principles of Morals* (Oxford: Clarendon Press, 1975), pp. 18–19.

[31] W. Blake, *Poems and Prophecies* (London: J. M. Dent & Sons, 1972), p. 3.

VI

The desires and perceptions of man untaught by anything but organs of sense, must be limited to objects of sense.

Conclusion

If it were not for the Poetic or Prophetic Character, the Philosophic and Experimental would soon be at the ratio of all things and stand still, unable to do other than repeat the same dull round over again.

12   Does Popper Explain Historical Explanation?

KENNETH MINOGUE

It is one of Karl Popper's great distinctions that he has an intense—some would say too intense—awareness of the history of philosophy within which he works. He knows not only its patterns, but also its comedies, and sometimes he plays rhetorically against their grain. He knows, for example, that the drive to consistency tends to turn philosophy into compositions of related doctrines, each seeming to involve the others. Religious belief, for example, tends to go with idealism and free will, religious scepticism with materialism and determinism. Popper does not believe in a religion, was for long some kind of a socialist, and takes his bearings from the philosophy of science. Aha! it seems we have located him. Here is a positivist, a materialist, probably a determinist. But of course he denies he is any of these things. Again, like many modern thinkers, he wants to extend scientific method not only to the social sciences but also to history. So far so familiar, until we discover that he regards nature as no less 'cloudy' than human societies.

No doubt he is also aware of the comedy whereby a philosopher who claims to have been able to dispense with some basic component of the systems we have inherited finds himself charged with harbouring its vestiges. Hegel

357

dissolved God into Geist but his followers thought he had only swept the concept under the carpet. Popper's 'God', as it were, was induction, and some notable lecturers in the present series have detected its presence, as indispensable, in Popper's own system.

Modern philosophy, which Popper once told me he regarded as rather pitiful, facilitates surprising doctrinal combinations because, especially in post-modernist circles, the consistency requirement has been greatly relaxed; or perhaps, has been found to be less coercive because modern philosophers have come to terms with the contingency of experience. But this loosening is not the reason Popper himself has been able to make his own unique philosophical pattern, for he insists that science and history work in the same way. He strongly affirms a unity of method and it is this doctrine on which I propose to focus.

I

Popper's unity of method is enormously far reaching. In *The Poverty of Historicism* it covers the natural and social sciences, and pushes out into politics in the form of the theory of piecemeal social engineering. As the rationality principle came to be developed in Popperian circles, it revealed itself unmistakably as a prescription for dealing with reality itself. Human life in these terms ideally consists in facing and responding to a set of problems by conjecturing solutions and testing them by the process of trial and error. Such a formula covers the detective and the entrepreneur no less than the physicist and the economist, revealing I suggest that the motor of the argument is a basic practical belief that we ought to be critical and rational in responding to the world. In time, the method came to be embedded in the universe itself as Popper increasingly incorporated evolution into his own system. The world works not by incremental growth, but by trial and error.

Incorporated in Popper's philosophy, then, was a message: that critical rationalists were both morally and in practice superior to those who lived by ritual and routine. The famous distinction between open and closed societies contrasted ideal forms of this distinction, and was, of course, immensely successful in formalising the conflict between liberal democracy and totalitarianism which was such a feature of his lifetime.

Method in Popper's thought was thus highly resonant. It derived its specificity from science, and came in time to cover the explanation of the entire human world. A single comprehensive method is assumed throughout Popper's work from the beginning. It is a theme concisely treated in the essay 'On the Theory of the Objective Mind' in which Popper posits World 3, a realm of ideas and implications logically independent of the subjective environment in which we entertain such ideas. In Section 11 of that essay he denies the view, represented by Dilthey and Collingwood, that understanding the human world differs from nature in any essential way. It must be said that the 'similarities' he suggests between what are sometimes taken as separate realms are rather knockabout stuff: 'Labouring the difference between science and the humanities has long been a fashion and has now become a bore' he tells us, impatient to get on.[1] He does pause, however, to supply four reasons why there is no difference. They are rather odd.

The first is that just as we understand other people, to the extent that we do, because of our shared humanity, so we may understand nature because we are part of it. Unity of method is a magnet to which bits of unity of substance have clearly begun to cling. The odd thing is that precisely this classical assumption—that man shares a common form of reason with nature—had long been thought responsible for the postulation of final causes, from which science with difficulty freed itself only in early modern times. The experimental method has commonly been traced to the mediaeval doctrine that human beings are *outside* nature, and therefore, not being able to understand it from the inside, must put it to Baconian torture.

Popper's second reason is merely a more precise specification of the first: 'As we understand men in virtue of some rationality of their thoughts and actions, so we may understand the laws of nature because of some kind of rationality or understandable necessity inherent in them.' What has traditionally distinguished understanding from explanation in science, however, is not some sort of 'understandable necessity' but the presence of meaning. We do not usually see changes in plants or atoms as revealing meaning. Popper has generalized scientific method to a level at which it can cover any kind of rational attitude at all. He has not

---

[1] 'On the Theory of the Objective Mind', *OK*, p. 185.

incorporated the element of meaningfulness which marks off the human world.

His third point is not one I understand. It is that we understand nature like a work of art—as a creation—a point, apparently, on which the non-believer Popper is prepared to countenance in some degree Einstein's references to God.

And finally we are told that method in science and the study of the humanities is the same because in both cases it fails to achieve full understanding: a *non-sequitur*. All understanding is no doubt incomplete, but propensity to failure hardly establishes a specific similarity.

There is, then, a single method which is the key to understanding and mastering the world. World 3 was a late flowering implication of this doctrine, a vast storehouse of problems and implications contingently related to the mental experiences of World 2. Popper's World 3 is a form of Platonism to which philosophers sometimes have recourse in solving difficult problems, and it is significant for my argument because it resembles Collingwood's solution to a very similar problem: namely, his doctrine of re-enactment, the process of understanding which allows the historian to grasp past thought. Popper is clearly aware of the affinity, and he shadows Collingwood as he develops his argument, only to represent his own view as 'diametrically opposed.' The 'diametric' bit is that Popper thinks that the problem situation of a historical personage, being part of World 3, is objective and quite distinct from the historian's mental experiences in World 2, while Collingwood takes 'the historian's mental process of re-enactment, the sympathetic repetition of the original experience' as the key to understanding.[2]

It is an interesting question, however, quite what the difference on this point actually is between the two. One difference there certainly is: Collingwood rejects the ontology of World 3, remarking of any thought (such as an argument of Plato) that 'without some appropriate context [of mind] it [i.e. a thought] could never exist'.[3] But Collingwood in his theory of re-enactment does very carefully distinguish the thought as a re-enactable entity in some respect outside time from the subjective flow of mental events in which it is certainly embedded.[4] The process of understanding by which the historian grasps the thought of a past actor is a genuine process of understanding. It must grasp, as it were, the objective element of a thought, and is not tied to the 'sympathy' Popper adds to the mix. It is this objectivity which makes it the same thought. My view is that

[2] It is odd that Popper should choose this principle as the way of distinguishing himself from Collingwood, since in PH, p. 138, he had stressed the affinity between social and natural science by suggesting that the physicist 'quite often uses some kind of sympathetic imagination or intuition which may easily make him feel that he is intimately acquainted with even the "inside of the atoms"—with even their whims and prejudices.' He adds, of course, that this intuition is the physicist's private affair.

[3] R. G. Collingwood, The Idea of History (Oxford: Clarendon Press, 1946), p. 301.

[4] Thus: 'It is not only the object of thought that somehow stands outside time; the act of thought does so too: in this sense at least, that one and the same act of thought may endure through a lapse of time and revive after a time when it has been in abeyance' (Idea of History, p. 287).

Collingwood worked very hard, and carefully, on this problem, and came (the ontology of World 3 apart) to very similar conclusions to those of Popper. That they are similar is a point supported, in my view, by the fact that both of them tended to choose their examples very carefully in order to reinforce the plausibility of positing ideas as timeless and more or less independent of their psychological context. Collingwood's standard examples deal with dealing with Euclid and Plato,[5] Popper with Galileo's theory of tidal movement.[6] Reconstructing the problem situation of Charlotte Corday, or Napoleon at Waterloo would undoubtedly be a much messier business, though Collingwood's discussion of Nelson's remark 'in honour I won them, in honour I will die with them' shows what can be done.[7]

II

Popper's unity of method takes off from a modified view of the way scientists conduct themselves, incorporates the social sciences, and moves on to take over history, or part of history. Historical explanation, Popper argues, is causal explanation, and cause and effect must, for logical reasons, be mediated by a law. This issue became a notorious battlefield in mid-century philosophy, focused around what was called (though not by Popper) the 'Popper–Hempel' theory of historical explanation. Popper calls it 'his' theory, as expressed by Hempel,[8] but it was

[5] *Idea of History*, Part V, 4, pp. 282ff.   [6] *OK*, pp. 170ff.
[7] R. G. Collingwood, *An Autobiography* (Oxford: Clarendon, 1939), p. 112.
[8] *PH*, p. 144n.

Hempel who actually got down to the underlabour of working out what it might actually mean for history. Controversy long raged (and indeed continues to rage) over the question of how tight such a law would have to be.

Popper's own treatment of the issue can best be described as 'insouciant'. The laws which 'glue' historical events causally together, he tells us, are often trivial; in history our interest is in the particular. For example, we do not need a law of the combustibility of human beings to explain to us why Giordano Bruno died in the flames in Rome in 1600.[9] We can take it for granted. This example is, of course, a natural not an historical connection, and simply misses the point. Next he attempts to exemplify these implicit trivial laws, Popper supplies us with a list of examples,[10] about which Alan Donagan has rightly observed that they are all either vacuous or false, and adds: 'If Popper... had known of even one "correctly enunciated" sociological law that was "wholly consonant with current evidence," I cannot believe that [he] would have refrained from stating it. I infer that [he knows] of no such law.'[11] Davidson, in discussing Hempel's theory of action, is happy to agree that causal explanations involve a law relating events under a description, but is careful to circumscribe its range:

> An explanation unconditionally predicts what it explains (in the sense that the sentence to be explained can be

[9] *PH*, p. 145.   [10] *PH*, p. 62.

[11] Alan Donagan, 'The Popper–Hempel Theory Reconsidered', in William Dray (ed.), *Philosophical Analysis and History* (London: Harper & Row, 1966), p. 145.

deduced from the law and the statement of antecedent conditions), and conditionally predicts endless further things. So by asking what an explanation of a particular event conditionally predicts we learn what sort of law is involved.[12]

The sort of law he concludes is involved turns out to relate to beliefs and dispositions, and may tell us quite a lot about the actor, but it will tell us very little about mankind. As he goes on (using his example of a soluble sugar cube), to say:

Explaining why something dissolved by reference to its solubility is not high science, but it isn't empty either, for solubility implies not only a generalization, but also the existence of a causal factor which accounts for the disposition: there is something about a soluble cube of sugar that causes it to dissolve under certain conditions.[13]

Consider an event: 'Have dinner with me', I said to the girl at the table. She slapped my face.

No law or convention of a general kind could cover the intelligibility of this sequence. If a causal law, it would have to be something weak about the character of the girl herself. But do we need a *causal* connection at all?

III

Popper's thesis that history involves explanation in the same sense as science begins to crumble as soon as one considers

---

[12] Donald Davidson, 'Hempel on Explaining Action', in *Essays on Actions and Events* (Oxford: Clarendon Press, 1982), p. 273.

[13] Ibid., p. 274.

the character of the laws that might be needed to provide the 'glue' for the explanatory structure. But the widespread fascination with the logic of causality in this area has distracted attention from a more fundamental defect of Popper's takeover bid for history. The defect is that it fatally divides history into two unrelated parts: firstly, the red meat of real historical explanation supposedly revealed by bringing covering laws to the surface of attention, and second a certain amount of mere fat constituted of the reporting of facts which are 'accidental' or merely 'interesting'.[14] As he explained:

> One of the most important tasks [of the historian] is undoubtedly to describe interesting happenings in their peculiarity or uniqueness; that is to say, to include aspects which it does not attempt to explain causally, such as the "accidental" concurrence of causally related events. These two tasks of history, the disentanglement of causal threads and the description of the "accidental" manner in which these threads are interwoven, are both necessary; at one time an event may be considered as typical, i.e. from the standpoint of its causal explanation, and at another time as unique.[15]

In this way, a great deal of history is drummed out of intellectual respectability as lacking explanatory significance; it is merely the description of the unique and particular. John Watkins, who has expounded the causal theory of historical explanation in a Popperian idiom, expresses this

[14] *PH*, pp. 146–147.   [15] *PH*, p. 147.

assumption by arguing that if situational logic can be used to *explain* some historical events, 'then the area of the arbitrarily given, of sheer brute fact in history, although it can never be made to vanish, will have been significantly reduced'.[16] History is thus divided into respectable facts, married to laws, on the one hand, and 'brute facts' on the other.

It would seem that this view gives us a clue about the 'problem situation' which generates the Popper–Hempel–Watkins theory of history. It starts from the view that history is simply an assemblage of historical facts (William led an army to Britain in 1066, defeated Harold, marched on London, etc.). Low grade stuff, indeed; mere chronicle. Yet it is hard to deny that history does explain what happens in the past, and is a valid answer to questions like 'Why did William defeat Harold?' or 'How do we explain the outbreak of revolution in France?' And if history *explains*, and we know what explanation is, then it *must*, in a concealed way perhaps, exhibit the same logical structure with which we are familiar both in science and in the field of practice. *Ergo* laws, even if concealed, or ignored, because trivial. And this is plausible because historians themselves are always talking in causal language of one kind or another. Coming at history from this angle, Popper is a knight on a white horse, saving history from intellectual nullity.

The intellectual nullity of history however only arises if one assumes the Popperian thesis about explanation

---

[16] J. W. N. Watkins, 'Historical Explanation in the Social Sciences', *British Journal for the Philosophy of Science*, Vol. VIII, No. 30 (1957), p. 117.

in the first place. If history were such an unstable mixture of explainable and merely decorative facts, it would resemble curdled mayonnaise. It is hard to see how the two elements could fit together. Yet history, if not a seamless web, is certainly a story, and a story is not an unstable compound of necessary, causally determined events combined with accidental or decorative features. It has a continuity and coherence all its own.[17]

The problem, then, is to give an account of what makes a story persuasive, and the philosophy of science is a model likely to lead us in precisely the wrong direction. Popper's account of explanation requires a *necessary* connection between the events being described, and we have seen Watkins dismissing the non-necessary as being *arbitrary*. The point of a story, however, is that the actions described in it are neither necessary nor *arbitrary*. They are contingent. This means that the story itself imposes its own persuasiveness by a kind of intelligibility whose force in no way depends upon necessity. There are indeed occasions when we may respond to a narrative by disbelief, and it is on these occasions, and generally only on these occasions, that a narrator may remedy any defects of plausibility by generalising. W. B. Gallie compares following a story to understanding a game one is watching, in which someone who knows the game may explain a difficult passage to a new

---

[17] As A. C. Danto puts it: 'we read a narrative with the expectation that each thing mentioned is going to be important...' Arthur C. Danto, *Narration and Knowledge* (New York: Columbia University Press, 1985), pp. 355 (italics in text).

spectator by elucidating rules. But this activity is not (as with Popper–Hempel) at the heart of the matter; it is, rather, peripheral. As Gallie remarks:

> what appear to be explanatory sentences—or what might be taken to be such by overzealous logicians—can often perfectly well be replaced by a number of narrative sentences which no one would dream of regarding as explanations.[18]

Basically, then, narrative has a unified texture in which everything belonging to it is marked by *contingency*. The interest of a story lies in the fact that when A responds to what B has said and done, we are interested to discover what the response will be. If the response were strictly necessary, predictable in all its details, it would lack all suspense, all interest. And if it were merely 'sheer brute contingency' it would tell us nothing about either the person or the situation. It might, indeed, as Popper suggests in what we have quoted, tell us about typicalities. But with typicalities we are already over a shadow line into sociology.

IV

The doctrine of the unity of method reduces all materials to a single form. The *intelligibility* we find in a story must be assimilated to the *necessity* which relates events by scientific laws. Scientific theories may be tested by predictions, but we

---

[18] W. B. Gallie, *Philosophy and the Historical Understanding* (London: Chatto & Windus, 1964), p. 111.

cannot in a narrative stop, as it were, the narrative and attempt to *predict* (though we might be led to guess) what will happen next. In practical life, of course, we do make such guesses all the time—but we can only do so in precisely the general terms which have no status in history. History is not an account of events in general terms. There is no law such that, from knowing that a number of Romans feared for the Republic at the hands of Caesar, one might deduce the assassination of Caesar at the Senate House on the Ides of March.

Our situation has certain analogies with a contrast between Plato and Aristotle. Plato's version of the doctrine of unity of method led him to dismiss rhetoric merely as bad logic. For Aristotle, on the other hand, rhetoric was not indeed properly deductive, but was to be recognized as having a value and independent status of its own. The problem is, then, to find an account of the evident intelligibility we find in history which firstly does not reduce it to science, and secondly does not divide it into two areas with greatly differing values. One solution to this problem has been provided by Michael Oakeshott.

Oakeshott argues that the Popper–Hempel thesis begs the question of what is distinctive about historical understanding. Oakeshott supplies us instead with a careful, indeed fastidious, account of what is an event (it is an outcome) and how it is circumstantially related to other events in such a way that they become intelligible. History, understood as an intelligible composition of circumstances, is essentially descriptive—but not *merely* descriptive. This is description robustly capable of carrying a heavy load of inference. It tells us *what* happened, and when it has successfully done that, it has already told us (in Collingwood's formula) *why* it happened. The issue revolves, then, around the question of what happened. Popper's account takes it for granted that we know *what* happened: the question is, how do we explain it. But on Oakeshott's argument, by the time we know *what* happened (or think we do) the work of historical understanding has *already* been done. For the historian has created by inference out of the evidence available to him an account of the course of events such that in the first place, each event is evidenced, and secondly, the events fit together with the same intelligibility by which we understand a story. What is wrong with a covering law account of historical understanding is, then, that:

> what must be the main concern of an historical enquiry
> —to understand the character of a not yet understood
> passage of a past which has not survived—is dismissed as
> a nugatory engagement in favour of a design to raise the
> occurrence of an alleged already described and
> understood kind of happening from the status of a report
> to that of a retrodicted necessity.[19]

Popper's view of history is indeed remarkably simple. It seems to be a matter of whether a report corresponds to the fact, as revealed in the remark that

> Clearly, no historian will accept the evidence of
> documents uncritically. There are problems of
> genuineness, there are problems of bias, and there are

---

[19] *On History*, (Oxford, 1983), p. 181.

also such problems as the reconstruction of earlier sources.[20]

This is a good example of the naivety of the critical attitude, for it is to think that the historian's question is: should I believe my sources? The point is that historians are asking questions quite different from those which the sources might be thought capable of answering.

It might be possible, indeed, to take one or another such historical account, turn the events into abstractions, connect them causally and in this way create a historical sociology. No doubt this is possible. But it is not history, nor, in Oakeshott's view, would such an intellectual creation be *magnifique*. For what Popper is really talking about is a kind of historical sociology which turns historical events into abstractions and theorizes them. In rationalising the work of historians, the critical rationalist imagines he has created it. It is the fly on the axle wheel exulting in the dust it thinks it has created.

V

There does remain, however, a certain amount of common ground between Popper and some of his critics. It might perhaps be found in one way of explaining historical action which is common to both Popper and Collingwood. It consists in exploring the 'situational logic' of the historical actor. In *The Poverty of Historicism*, Popper tossed off this

[20] *CR*, p. 23.

phrase and did not stay for elaboration. Later, the notion turns up in his discussion of World 3 in his thesis 'that in all understanding, including the understanding of persons and their actions, and thus *in the understanding of history*, the analysis of third-world situations is our paramount task'.[21] As in our discussion of World 3, we find that there is a clear affinity between Popper and Collingwood, specifically with Collingwood's logic of question and answer,[22] at least as that doctrine applies to history. The major difference between them is that Collingwood's cognitive foundation is in history, which he takes to be the basic science of human nature, while Popper's is in science. It is this difference which accounts for what distinguishes their uses of a similar idea. Popper wants to explain actions by entering into the problem-situation of historical actors, a process which I take to be basically similar to Collingwood's re-enactment. A kind of merger between the two will be found in John Watkins's ambitious exploration of the rationality principle.[23]

Situational logic attempts to formalize the structure of action. It has been elaborated into an intellectually rigorous area of inquiry in modern decision theory. In decision theory, desires may generate a rational structure of preferences, and the preferences may be judged in terms of their

---

[21] *OK*, p. 167 (italics in text).

[22] Collingwood, *An Autobiography*, ch. 5.

[23] John Watkins, 'Imperfect Rationality', in R. Borger and F. Cioffi (eds), *Explanation in the Behavioural Sciences* (Cambridge University Press, 1970), p. 167.

likelihood, in order to yield 'expected utilities'. The reasons for acting in a particular way may generate implications which connect the actor and the action by some kind of hypothetical imperative—and the point is, indeed, to get some kind of necessity into the explanation. Here then might seem to be the location of causal connections which would allow Popper to get beyond contingency into the necessity he craves, and which he craves because he seeks to assimilate the wayward field of human action to this conception of explanation.

Rational choice theory, however, emerges from a merely formal structure of understanding, having little relation to the actual messiness of human conduct. What, for example, is the thing called the 'action'? To be understood, it must be described, and any action can be described in many ways, yielding very different judgments. ('I am doing the right thing!' may be met by 'You are betraying your country.') Don Quixote charged what he thought were a set of giants who turned out merely to be windmills, in a scene which has come to be emblematic of the complexities of the modern world. Again, are the desires and reasons by which we act fully available to us at the time of action? It is certainly true that we do have desires, and reasons, and we act. That is very far from being the end of the matter, and later we may realize that other considerations, hidden dispositions the significance of which only become clear in retrospect, were at work. Sometimes, indeed, we are clear about hidden reasons for action (malice, for example), but we keep such impulses to ourselves. And finally, there are those cases where all the reasons seem to indicate one specific action and I intend to perform it, but at the last moment, I do something quite different: such is often the aetiology both of folly and of self-sacrifice.

We need not labour the complexity thesis. All theoretical accounts of human action look thin beside our sense of the mysteries of conduct. One element that is highly relevant to our theme, however, is the moral question. And here we are not dealing with a desire at all. We are dealing, rather, with the issue of identity.

A moral identity is usually a background condition of action. It is often a prohibition blocking off a course of action one finds unthinkable—not stealing, murdering, lying, etc.—the violation of which would induce guilt, shame and self-disgust. The most conspicuous forms of moral identity are preserved in some sort of religious fluid, but they will be found, in often eccentric forms, among mafiosi, or in such cases as the widespread contempt among other criminals for sex offenders.

It is one central feature of what (to go along with Popper's term) we may call 'situational logic' that there is commonly a conflict between the instrumental considerations which guide us in the satisfaction of desire, and the 'intrinsic' considerations which guide us in preserving our moral identity. This is a rather pompous way of referring to the phenomena of temptation, and they are obviously very important in making sense of the human condition. In the phenomenon of temptation, a variety of mental and moral safeguards are under attack, and the reasoning powers of the tempted are notably enfeebled. What might keep us on the right track, if there is such a thing and we are kept on it, would be some entrenched conviction. Popper's critical rationalist, however, holds nothing sacred, including sacredness itself, and entrenched convictions must be subject to challenge and refutation. Where the temptation is serious, they are likely to be refuted very smartly indeed. Popper is thus a paradigm theorist of liberations—or of what is currently referred to by sociologists as 'detribalisation.'

The phenomena of moral integrity, by contrast, are a powerful barrier to making predictions about human life. No doubt the structure of desires in a society is immensely complicated and volatile, but one might in principle attempt to aggregate it and make a prediction about what it might generate. And this is indeed what can be done in the practical world by people ranging from actuaries to politicians. What prevents these predictions from being more than informed guesses is above all the fact that the way people act is unpredictably circumscribed by moral identities which, for all our continual discussion of issues of ethics and morality, flourish in the darkness of conscience.[24]

Let us add another complication, by asking whether it is plausible to see life as a succession of problems. This is Popper's view of the human condition: 'My thesis is that the main aim of all historical understanding is the hypothetical reconstruction of a historical *problem-situation* ... we can

---

[24] What account might one give, for example, of the recent case of a Maori woman who was hired as a contract killer, and who, after doing the job, returned to New Zealand, was overcome by religious conversion, and confessed, involving her principals in her downfall. Was her action rational, or irrational?

interpret an action as an attempt to solve a problem.'[25] The idea that life is just a succession of problems to be solved is a conception often derisively attributed to Americans by Europeans, and it is no doubt true that some people are more 'problem-oriented' than others. Intellectuals and the middle classes are sets of people who generally respond to life far more in this active, problem-solving way than do peasants or the natives of traditional societies. We may take life as a succession of problems, and no doubt everyone sometimes does; but we may *also* take it as a condition of things which must be accepted—indeed much of religion formalizes an attitude of graceful acceptance towards the world. But beyond this, the reality of life for most people most of the time is that it is a flow of action guided by habit, routine and principle, and it is only *within* this flow that problems may arise. It follows that in understanding some things—for example the conduct of foreign policy—we may well have to concern ourselves at times not with problems, but with the almost insensible flow of routines.

Human life in its complexity, then, may appear as a condition of things to be accepted, or as a flow of routine, and only in part—but the part that is most salient in our modern Western world—as also a set of problems to be solved. And this is important in determining just what the action to be explained actually is. How, for example, do we understand Sir Thomas More's conduct in refusing to recognize the religious Supremacy of Henry VIII? Since he

---

[25] *OK*, p. 170, p. 179.

ended on the block, we might well say that he spectacularly failed to solve his problem. On the other hand, in terms of his demeanour and his steadiness in holding to a chosen moral identity, we might well judge it, as later generations have, as a triumphant success. Is the hero of Hemingway's famous story *The Killers* being irrational in refusing to flee those he knows are coming to kill him? It is not merely that moral issues make this a difficult situation to construe. There are many cases in which we cannot properly describe what the action was until we know its consequences. Consider an entirely hypothetical prime minister facing an important vote, whose whips use rough tactics in getting dissident members of parliament on side. If he succeeds, and especially if he succeeds easily, he will be said to have 'over-reacted'. If he had left things to chance, and the motion had failed, he will be said to have been complacent. Popper can see two general ways of characterizing an action: as the agent sees it, and as (by conjecture) what the interpreter thinks are the realities of the situation. But human situations are no less complex than those visual puzzles that Wittgenstein discusses when treating the concept of 'seeing'.[26]

There is, of course, a ready and easy way to solve this difficulty. It consists in assimilating both what I have called 'moral identity' and also the common acceptance of one's current condition into the structure of desires. It will then be one condition of solving one's problem that the act

---

[26] Ludwig Wittgenstein, *Philosophical Investigations* (Oxford: Basil Blackwell, 1974), II, x.

must not only yield satisfaction but also sustain any desired moral identity.

This is an interesting, and I think significant possible development. It is unmistakably a 'degenerating problem shift' because its simplification obscures a vital distinction. It would involve stretching the concept of desire to cover all human action and (perhaps more to the point) inaction. The point of a desire, however, is to achieve satisfaction, and if it can be shown that a different course of action is more satisfying, then it would be rational to follow it. But a moral identity is not a bid for satisfaction, but a condition of enjoying satisfactions. Similarly, to see every action as the attempt to solve a problem is to stretch these simplicities to the point where they become a form of descriptive fundamentalism.

That this is a problem for Popperian ethics seems to me clear. It is clear, for example, in the fact that Popper chooses examples which do not raise any moral difficulties. So incidentally does John Watkins in his fascinating account of the failure of a British admiral to keep his fleet in being. The problems are entirely intellectual and technical. And this raises the more general question: What *is* Popper's moral theory?

It is notably thin. It often consists of vague talk about 'values' and their place in problem solving.[27] It rests upon a dualism of facts and values in which values are decisions we take and norms we choose to apply. In a wider

[27] *UQ*, pp. 193–196.

context, we know where Popper stands. He is for the Open Society against the Closed; he deplores Utopian Social Engineering, and thinks that politicians should proceed with caution—piecemeal, in fact. He is in favour of critical thought rather than unexamined dogma. And he thinks that the responsibility of governments is to facilitate the happiness of their subjects—though his admirable caution is that they can best do this by removing evils than by positive attempts at good.

This is what the Americans might call a 'motherhood' programme among the educated Western classes of our time. None of us will fail to salute such flags. But it is so conventional, and so tied up with political issues, that one cannot but suspect that this is an area of trouble for Popperian philosophy. Perhaps the explanation is that, being self-described as the happiest of men, he was quite content to spend his life solving philosophical problems, except for his understandable involvement with the totalitarian evils of our century.

But it might also be that there is not much moral substance in Popper's thought at all. He is a cheerful rationalist seldom troubled by any very deep moral perplexities. He knows, of course, how to respond to Auschwitz, but his ideas about human life all seem to revolve around happiness, and science is valuable because it conduces to progress as the enjoyment of happiness. There is no doubt that religious dogmas stand directly opposed to the infinite revisability of the basic message of critical rationalism. If everything may be understood as a mode of religiosity, we might construct from Popper's writings the message which has made him so powerful a guru in our time: Don't be afraid to revise any conviction which stands in the way of your happiness.

V

Let me now move to a conclusion. We have been considering Popper's thesis that there is a method, basically a scientific method, which underlies our understanding not only of nature but also of the human world. Popper was a kind of academic imperialist, but a benign one. His method allowed, in a way, for uniqueness, particularity, indeterminacy and other features of the human world which were often thought to be in danger from the simplicities of positivism. Popper insisted that he was certainly not a positivist.

He sought to replace contingency and the intelligibility found in narratives by deductive structures generating, in so far as possible, causal necessity. In one version of Popper's rationality principle, Watkins tells us that 'to provide a conjectural explanation of a past action is to postulate a decision-scheme which has a practical conclusion of which that action could be the natural outcome'.[28] This is clearly a useful scheme in dealing with the type of situation which concerns Watkins in his account of the rationality principle —a situation, that is to say, in which the explanation reveals that there were failures of communication. In cases where the action is more ambiguous, and especially where there is evidently a significant moral dimension, the schema either

[28] 'Imperfect Rationality', p. 209.

merely formalizes what is clear enough without it[29] or alternatively imposes a misleadingly simple idea of necessity upon it. As we saw in discussing Donagan, the laws which might supply the necessity can only be formulated in ways that turn out to be false.

Standing back from this argument, we might suggest that our understanding of the world flows downward from its source in our experience towards the actions and propositions we consciously produce. Science is an impressive form of explanation at this level of explicitness, and Popper is totally at home with it. But science, like the rest of our knowledge, depends upon processes of understanding which are much more difficult to grasp. As is clear from his famous schema about ideas[30] Popper prefers to think about truth than to think about meaning. It has to be said that this bent leaves him more than more philosophers at the mercy of the dominant fashions of his time, both morally and intellectually.

---

[29] Watkins is clear, for example, that the actual explanation of how the naval disaster occurred had first been given by a naval writer in *The Times* a month after the disaster happened in 1893 ('Imperfect Rationality', p. 214).

[30] 'On the Sources of Knowledge and of Ignorance', in *CR*, p. 19.

13   The Grounds for Anti-Historicism

GRAHAM MACDONALD

Introduction

In his seminal *The Poverty of Historicism*[1] (hereafter *PH*) Sir Karl Popper deployed a number of arguments to prick the pretensions of those who thought that they were, or could come to be, in possession of knowledge of the (social) future. These 'historicists' assumed that they could lay bare the law of evolution of a society, and that their possession of knowledge of such a law justified (large-scale) political action which had the aim of removing obstacles to the progress of history. In arguing against historicism Popper was clearly motivated by his interest in removing the intellectual backing for such revolutionary political practice.[2] My first reading of *PH* was in the company of people who were extremely dismissive of

---

[1] The seminal character of Popper's argument is testified to by Isaiah Berlin: "[Popper] exposed 'historicism' with such force and precision, and made so clear its incompatibility with any kind of scientific empiricism, that there is no further excuse for confounding the two" (I. Berlin, *Historical Inevitability*, Oxford University Press, 1954, pp. 10–11).

[2] 'Popper's most passionage disagreements with historicism are over its implications for practice' (A. Donagan Popper's Examination of Historicism, in *The Philosophy of Karl Popper*, P. A. Schilpp (ed.), Open Court, Illinois, 1974, p. 914). Popper also decried historicist quietism, a doctrine which claimed the futility of any intervention in the social arena.

the anti-revolutionary message, and who tended to argue that if that was the conclusion of Popper's theoretical argument, then obviously the argument was flawed. Within their context, that of the implementation of apartheid policy in South Africa, there was much to be said for this attitude. There is no doubt that Popper's message was insufficiently contextualised, or rather that he did not signpost very clearly whether he intended the anti-revolutionary political prescription to have limited or universal application. In this paper I want to reconsider some of these issues, particularly whether the truth of anti-historicism, in the sense intended by Popper, has such conservative consequences for political action. In so doing I wish to proceed in the best Popperian tradition by testing the arguments for 'anti-historicism' against the strongest case that can be made for historicism. Much will depend upon how historicism is characterised, so in the first section I provide a thumbnail version of the account offered by Popper. In the process I divest the original account of elements that are extraneous to the main theme, and also try to lessen the tension between disparate parts of the original. The second section constructs two differing accounts of theories of history which would appear to contradict some of the conclusions of the anti-historicist arguments. Section three adjudicates. Section four briefly considers the political implications of the verdict.

1.   Popper's Version

Historicism is presented as an amalgam of differing tendencies and assumptions, some philosophical, others more semi-empirical. There are some components which are anti-naturalistic, others pro-naturalistic. Welding the two together is not an easy task, so I will simply outline some of the anti-naturalistic assumptions and then proceed to drop those which are not directly relevant to the main thrust of the anti-historicist argument. The pro-naturalistic suppositions of historicism are more interesting partly because the tenor of present times is pro-naturalist, partly because Popper had more sympathy with the broad outlines of the naturalism espoused, differing mainly in the details.

One form anti-naturalism about any area of inquiry can take is to deny that the subject matter of the inquiry can be treated in a suitably scientific manner.[3] The anti-naturalistic arguments of historicism stress the particularity of historical phenomena, thus implying that there can be no general treatment of historical change, no context-independent theory which would be comparable, say, to a physicist's theory about physical change. Some of these implications depend upon the idea that history is essentially *holistic*, that in dealing with human action it deals with intentional, hence meaningful, phenomena whose meaning is tied to the practices and customs of the local community. The meaning of a particular action, and a proper understanding of its consequences, can only proceed by situating

---

[3] Only one form, because naturalism is not just one doctrine. For further discussion, see Graham Macdonald, 'The Nature of Naturalism', *The Aristotelian Society*, Supp. Vol. **LXVI** (1992), pp. 225–44, and Tom Baldwin, 'Two Types of Naturalism', *Proceedings of the British Academy*, **80** (1993), pp. 171–99.

that action in a significance-producing 'field' of other actions and intentions. The action cannot be separated from its social context without its significance changing, and so the atomistic, causal-explanatory methods of the natural sciences are unfitted to this subject. In particular the quantitative description and analysis typical of physical science will be inapplicable to historical phenomena. On this view it is impossible to explain historical change causally, or at least impossible to do so with satisfactory precision. Instead of trying to ape these methods of physics, so the argument goes, one should aim for a different type of explanation of social phenomena, one which involves an essentially 'intuitive' understanding.

Now these claims are no doubt very important in assessing the right way to go about studying social change, but they are slightly peculiar in the context of an argument about the inappropriateness of claims to knowing a theory of historical change. The anti-naturalist, as sketched above, seems to be set upon agreeing that such knowledge is highly unlikely, not to say impossible. After all, if the subject matter is context-sensitive in the manner suggested, then it is implausible that there could be any such thing as knowledge of the laws of historical change, where these cover different historical periods and cultures. It is only by saddling this kind of anti-naturalist with further claims, such as the 'intuitive' understanding providing the interpreter of history with special insight into the historical potential of a particular society, or intuiting the 'spirit of the age', that Popper is able to make a case for anti-naturalistic historicism. Popper's treatment of anti-naturalistic historicism has been ably discussed elsewhere[4] so I will not pursue it here, particularly as some of Popper's reservations about it stem from his concern with the testability of the claims made by such anti-naturalists. This concern (about testability) manifests itself again in the argument against naturalistic historicism, and it flows from Popper's general philosophy of science. It is to this background that we must now turn.

Naturalism here is the doctrine that the social sciences do, or should, follow the methods and style of explanation (causal explanation) of the natural sciences. It contrasts with 'pluralism', the view that the subject matter of the different sciences is so diverse that a different method and/or explanatory style should be adopted. Popper was a unifier insofar as he subscribed to the belief that science is methodologically unified, and so agreed with this naturalistic aim, disagreeing only with some characterizations of the method invoked. However he also thought that there were important differences between natural and social science. In particular he thought that insofar as social agents acted rationally social scientists could help themselves to a method of 'rational reconstruction' in their generation of hypotheses concerning the intentions and motivations of such agents. This difference in subject matter produced by agent knowledge and rationality will become important in what follows, but for the moment I want to concentrate on what emerges

---

[4] See Peter Winch, 'Popper and Scientific Method in the Social Sciences', in *The Philosophy of Karl Popper*, P. A. Schilpp (ed.), pp. 889–904, and Anthony O'Hear, *Karl Popper* (London: Routledge and Kegan Paul, 1974), pp. 153–70.

from the supposed methodological unity of science. For Popper, the central tenet of the unity of science was that *testing* of hypotheses was always to be conducted in the same manner as that of the natural scientist. Testing hypotheses was to be carried out by determined attempt to falsify those hypotheses, attempts that were facilitated both by the adoption (for testing) of a 'conjecture' with the highest information-content and the subsequent rigour of the (empirical) procedure for falsifying that conjecture.

Why is 'attempted falsification' the hallmark of scientific method? Within the broad outline of his hypothetico-deductive approach to scientific explanation and testing it is important to stress the anti-inductivism and fallibilism which fuelled Popper's falsifiability criterion of scientific status. The anti-inductivism expressed a deep suspicion of claims to knowledge, whether these be within natural or social science. Popper thought that an inductive approach to testing involved an attempt to prove the correctness of the hypothesis being examined. Given that such confirmations could always be found, an unhealthy dogmatism resulted. Dogmatic claims to knowledge were, for Popper, the antithesis of a scientific approach, which should always be characterized by an alertness to the possibility of our most cherished beliefs being mistaken. This keen awareness of our fallibility generated the caution with which Popper approached claims to theoretical knowledge. Given that our empirical theories could never be proven beyond doubt, we should pursue the policy of rigorously testing hypotheses in order to falsify them. Falsification had the merit of being based on a deductively valid principle, and so should be the foundation of a unified method for all the sciences. The rigorous testing, which was to be the hallmark of good science, had to involve the scientist formulating a hypothesis which was as contentful as possible (i.e. ruled out more possible states of affairs than rival hypotheses), and then setting up experiments designed to falsify the hypothesis.

The requirement of maximum content for a hypothesis, consistent with it not having been falsified, is just the requirement that scientists produce bold, risky hypotheses for testing, and so leads directly to Popperian assertions about the revolutionary nature of the best science. Caution about truth-claims leads to boldness in disputing such claims. A methodology designed for scientific progress thus seems to consist in making mistakes as fast as possible. Consequently the best scientists are depicted as heroic risk takers, prepared to sacrifice cherished projects in the face of falsifying evidence produced by themselves. This demand for maximum content leads to the need for hypotheses to be formulated with maximum precision. Vagueness in both the hypothesis and the formulation of the 'falsifying' experiment has to be rejected because it makes falsification more difficult. Dogmatism is to be avoided by making falsification as easy as possible. A further hedge against dogmatism is provided by Popper's insistence on the institutional character of science, that science must be pursued in a social context dominated by the spirit of free critical inquiry. Within such a context any errors overlooking by some practitioners will have a good chance of being spotted by an alternative research team. Truth is best pursued by democratic means.

The combination of praise for a revolutionary scientific methodology and the naturalistic assertion that such a method was essential for *any* science has led some to think that Popper was being inconsistent in his rejection of historicism. After all, historicists made 'bold' theoretical generalizations about the course that history must take, and were also likely, so Popper claimed, to be revolutionaries, political activists wanting to bring about large-scale social change. However there are two main lines of attack deployed in the rejection of this charge of inconsistency, the first being motivated by exactly the same set of concerns that generated the revolutionary methodology: testability is to be ensured by precise prediction and careful testing. Call this the methodological argument. The boldness of the historists' hypotheses, according to Popper, consisted in a *disregard* for their testability. Historicists freely admitted that their predictions could not be as precise as those of physics, and seemed to glory in the freedom this gave them. They were wont to indulge in large-scale forecasts based on evidence culled primarily from history. Sociology was therefore thought to be theoretical history, where this involved the study of the operation of historical forces. Such a procedure embodied the anti-critical attitude which Popper felt to be the hallmark of bad science. The largeness of scale meant that there was simply too much going on for careful critical analysis. So many factors were likely to be involved in any major historical process that studying each of them to isolate the contribution each made to the overall effect was thought to be impossible. From this purely *methodological* concern, then, Popper drew the conclusion that best social science was to be done by the *method* of 'piecemeal engineering'. This method consists in introducing small-scale changes in order to monitor their effects, with the social scientists adopting the appropriate attitude of wanting to learn from their mistakes. Rigorous testing in both natural and social domains required this watchful tracking of cause and effect, a tracking best accomplished in controlled conditions. The closest one could get to this in the social domain was the monitoring of the effects of the small 'adjustments' introduced into the social scene. The methodological argument concludes that large-scale forecast typical of historicist social science is to be rejected on purely methodological grounds, grounds that cover both natural and social research.

There is a second form of argument to the same conclusion in *PH*, one which emphasises differences between the social and the natural testing procedures. The difference is that in the former case testing involves humans, so social experiments have consequences, measured in terms of human happiness and misery, that are not part of the natural scientific domain.[5] This is the misery argument. The potential for an increase of human misery caused by failed (or, indeed, successful) social experiments particularly troubled Popper. Given this potential a circumspect strategy, one which was prudent about social interference, was

[5] The interaction between the two domains obviously makes for a more complicated picture. Some natural scientific experiments have consequences for human misery, such as those involving testing the atomic bomb. Presumably Popper would have been as much against these as he was against large-scale social experimentation.

desirable. Avoiding dogmatism in this context meant that one should recognise the impossibility of being able to estimate precisely what the consequences of any social experiment might be. Provided that this is impossible one should avoid taking the risk of causing untold misery, and so should avoid experimentation which could produce irreversible consequences. Again small-scale social change is preferred, but this time because of a difference between the study of the human and non-human world.

The question immediately arises as to why it is impossible to have knowledge of future consequences of one's actions. Is this limitation on knowledge restricted to social consequences, or is it just part of Popper's general scepticism concerning our knowledge of the future, part of his anti-inductivism? It is, I suspect, a bit of both. There is a general thesis about the uncertainty of all empirical predictions which is supplemented by considerations concerning the 'unintended consequences' of social actions. This supplementary bit, however, is difficult to understand unless it is just an aspect of the general thesis. Otherwise more needs to be said as to why unintended consequences are especially problematic. The reason for this is that Popper identifies unintended consequences with unknown consequences. Once the two are separated it is apparent that there may be foreseen but unintended consequences of social actions, so their being unintended does not preclude knowledge of their consequences. It is unforeseen consequences that are the problem, and this just is the problem of lack of knowledge of consequences with which we started. Ignorance plus the potential for immiseration plus negative utilitarianism[6] produces the requirement (so the argument goes) that one's social changes be minimal or easily reversible.

Now if the whole brunt of the ignorance claim was to be carried by a general scepticism with regard to knowledge of the future, then Popper's methodological advice would not make sense. Following the advice to 'only make small-scale social changes' where this means 'changes whose consequences are reversible' requires some knowledge of which changes are small-scale in this sense. That is, it requires knowledge of the reversibility of the consequences of those changes, and this is knowledge of the future. Popper's argument threatens to be too powerful, undermining any connection between ignorance and limited social experimentation. We will return to this in section three.

There is a special 'social' route to ignorance which Popper puts forward in the Preface to the English edition of PH. He claims it is a refutation of historicism: '*I have shown that, for strictly logical reasons, it is impossible for us to predict the future course of history*' (*PH*, p. v.) The argument ('the knowledge argument') is very simple:

1. The course of human history is strongly influenced by the growth of human knowledge.

---

[6] Negative utilitarianism maintains that avoiding pain is better than pursuing happiness. It is needed in this context because by itself uncertainty about our social future could not produce a preference for small-scale change. If the pursuit of happiness outranked the avoidance of pain we might settle on a risk-taking strategy.

2. We cannot predict (by any rational methods) the future growth of scientific knowledge.
3. Therefore we cannot predict the future course of human history.
4. We must therefore reject the possibility of theoretical history, where the theory would serve as the basis for historical prediction.
5. So the fundamental aim of historicism is misconceived.

This 'refutation' depends crucially on the first premise and the specific interpretation given to the second premise, issues to which we return in section two. For the moment it is important to note what place this proof has within the overall anti-historicist argument. The thrust of Popper's argument is to prick the predictive pretensions of historicists *in order to deny the legitimacy of revolutionary social change.* What we have seen is that there is one strand of argumentation, the methodological argument, which makes a case for the desirability of small-scale change on the grounds of the need to track the effects of certain causes. This in itself says nothing about the need for these changes to be reversible. Within this context the 'small' in 'small-scale' means something like 'discriminable from other causes'. The difference between this methodological connotation of small-scale and the later anti-revolutionary connotation is best exemplified by imagining a society with a fairly simple social structure. Within this context it may well be very easy to trace the effect of some changes which have a devastating impact on the people in the society. For methodological purposes the trackability of the effects could very well make such an experiment *cognitively* valuable. Even in a complex industrial society it may well be an easy matter to introduce a destabilising cause (a new virus, perhaps) the effects of which, whilst hugely immiserating, are easily traceable. It is primarily the misery argument which bears the brunt of the anti-revolutionary message, and we have seen that this is supplemented by a claim that we are ignorant about future consequences, and so cannot estimate amounts of future misery. It had been suggested that too much ignorance is liable to be self-destructive for Popper's purposes, so what is needed is a special social route to ignorance. It is in this context that the above proof is to be placed. Given that it is a proof which purports to eliminate the possibility of a theory of history, it will be valuable to look at two recent attempts at such a theory in order both to appraise them in the light of Popper's strictures and to test the truth of the premises in the proof.

2. Two Attempts to Construct a Theory of History

Case A

One of the main targets of the anti-historicist attack was Marxist theory of history, as it was primarily this theory which Popper saw as 'legitimising' revolutionary demands for large-scale social change. What can contemporary Marxists say in the face of the above 'proof' that their project was doomed? One obvious line of attack for historical *materialism* would be to deny the first premise, perhaps

on the grounds that it begs the question against a materialism that is opposed to idealism, where this idealism is
supposed to assert just such an influence of ideas on the
course of history. The problem with this is its sheer
implausibility; it appears to be obvious that (some) ideas,
in the form of technological and scientific knowledge, have
been crucial causal factors in the development of some
societies. It is also now unfashionable to interpret Marx's
materialism in this fairly literal way, as positing as the
primary determinant of historical change some kind of
matter, where this is understood in an ontological manner.[7]
Some may have thought that a connection could be made
between this ontological materialism and Marx's materialism via the notion of technological determinism, where the
technology component was seen as the material embodiment of certain ideas. However this is again implausible, as
the material embodiment sometimes depended for its
embodiment on certain scientific theories, so the primary
influence in these cases is again scientific knowledge. One
of the most influential of current accounts of Marx, that of
G. A. Cohen,[8] interprets 'material', as signifying 'non-
social'. This account also makes room for the influence of
ideas on historical change. Described as briefly as possible,

---

[7] This ontological interpretation of historical materialism would align it
with materialisms such as mind-brain identity theories. Such views
seem far removed from Marx's thoughts.

[8] See G. A. Cohen, *Karl Marx's Theory of History* (Oxford University
Press, 1978). What follows is the briefest possible summary of a
complex account.

Cohen makes out a case for a Marxist theory of history which makes technological change, or, more precisely, changes in the forces of production, the motor of historical change. The forces of production form the material (non-social) base of the rest of the social structure, comprising the relations of production (economic relations) and the so-called superstructure (principally the political and legal systems). Humans have a fundamental motivating interest in increasing the productive capacity of the forces of production, and so technological change in the direction of increased productive capacity is to be welcomed. The economic structure functions to facilitate this growth in productive potential, with the superstructure functioning to support the economic structure in this role. Changes in technology may eventually produce a situation where what once was functional for the growth of the productive forces, say the economic relations typical of a feudal society, are no longer functional, and so restrict future growth. In this situation the economic relations will have to give way to be replaced by another set of economic relations which will favour enhanced productivity. This will also induce changes in the legal and political systems, as they will need to function to stabilise the new economics relations. When all of this happens a social revolution has occurred, and Marx's theory is a theory about the nature of such revolutionary change.

What is relevant to our concerns is the role played by knowledge, or ideas in general, in the account of social change, for if Popper is right our inability to predict future knowledge rules out the viability of any theory of history which makes such knowledge causally effective in producing historical change. On Cohen's account such causally effective ideas are ubiquitous. In the first place the forces of production are characterised as consisting in, amongst other things, raw material, labour power, machinery, and such knowledge which is relevant to increasing the productive power of the forces of production. This could be scientific knowledge which feeds into technological change, thus enhancing production. It is not only in the invention of new productive forces that knowledge will be effective. In addition our theories as to how economies best function, and our political and legal systems, will utilise ideas which facilitate the realization of productive potential. Allowing theories and ideas such a central role appears to make this interpretation of Marx hugely susceptible to the Popperian argument.[9] Future knowledge cannot presently be predicted, so how can there be a theory which purports to tell us what future change will be like? Before answering this question I want to consider a rather different, non-Marxist, theory of history which specifically addresses the problem of revolutionary change.

---

[9] Popper himself used other arguments to reject Marxist theory. One which is relevant here arises from his methodological prescription that falsified theories ought to be rejected. Given that the revolutionary predictions of Marxism turned out to be false, Marxist theory of history ought not to be considered a 'live' theory. Appraising this objection would take us far afield into the Lakatosian territory of positive and negative research programmes. General problems of prediction are discussed in the text below.

Case B

In a remarkable survey of a number of different social transformations Jack Goldstone has produced a theory of social revolution which depends crucially on the role of demographic change, in particular unusually high rates of growth of populations.[10] A rapidly increasing population causes both inflation and the potential for mass mobilization (due to an increase in urban population, a younger age structure, and the decline in real wages brought about by inflation); in addition the growth in population fuels élite competition (not enough positions of prestige for enlarged next generation of élites). Inflation exacerbates élite competition whilst at the same time producing state fiscal distress (increased costs, such as army wages, outstripping ability to raise taxes to pay for them). The combination of state fiscal distress, élite competition, and mass mobilization causes political stress which, if sufficiently high, will result in revolution (given the availability of a revolutionary ideology). Goldstone defines the *political stress indicator (psi)* as: fiscal distress × élite competition × mass mobilization potential, and shows that in the case of the English revolution in the early part of the seventeenth century the *psi* was very high. 'The constitutional struggles and religious conflicts of the times were underpinned by an extraordinary combination of decaying finances, rising élite mobility and competition, and falling real wages, all propelled by rising population. This

[10] See Jack Goldstone, *Revolution and Rebellion in the Early Modern World* (Berkeley: University of California Press, 1991).

combination of ills reached a peak in the middle third of the century unmatched at any time throughout early modern English history... The *psi* function clearly differentiates the mid-1600s from the preceding and succeeding centuries as a period of exceptional predisposition to state breakdown.'[11] Goldstone tests this model against the examples of the French revolution, crises in the Ottoman Empire (especially Anatolia and the Balkans), and the Ming–Quing transition in China, arguing that in each case his hypothesis is supported by the evidence. For example, in the Ottoman Empire Anatolia and the Balkans displayed instability at different times. Anatolia experienced a huge crisis in the seventeenth century, the Balkans being particularly rebellious in the nineteenth century. In both cases population growth, with its attendant fiscal distress, élite competition, and potential for mass mobilization, preceded the crises.

3.   Predicting History

So are Cohen's Marxist and Goldstone's demographic hypotheses examples of historicist theories that escape Popper's strictures? Are they theories of history which serve as the basis for prediction? If so, how do they avoid the knowledge argument? On Cohen's account Marxism would be weakly predictive. It predicts that normally[12] societies will

---

[11] Goldstone, *Revolution and Rebellion*, p. 145.

[12] A 'normalcy condition' seems to be a normal feature of explanation in the special sciences. The way Cohen puts it for the Marxist case is: if there were to be a meteor whose impact on earth destroyed all human

maximize their productive capacity by increasing the productive power of the forces of production; that in doing this
they will employ relations of production which will facilitate
the growth in productive power of the productive forces so
that at some stage the relations of production will no longer
be able to function to this effect because of the changes
brought about at the level of productive forces; and that
when this happens the relations of production will be
replaced by a more efficient set of relations, and a new
superstructure to legitimate these new relations will have
to be put in place. A revolution will occur. More specific
predictions may be possible given the more detailed knowledge we may have concerning the nature of the mode of
production which dominates in a particular society. So,
Cohen speculates (or predicts), in a capitalist mode of production, if a choice is to be made between sharing work (and
so working shorter hours so there is more work to go round)
and making some work full time (and over-time) whilst
leaving others unemployed, then the choice will be determined by which option is the more profitable for employers
(generally the second option).[13]

society, it would not invalidate a theory which claimed that in normal
(meteor-free) conditions, societies tend to develop their productive
forces. The case for such normalcy conditions in biology is
persuasively argued by Ruth Garrett Millikan in 'Thoughts Without
Laws', *Philosophical Review* **95**, no. 1 (1986), pp. 47–80. An interesting
feature of Millikan's case is that she denies that biology is predictive.

[13] Those Marxist predictions which stem from the labour theory of value
(e.g. that there will be a tendency for the rate of profit to decline) are
discarded by Cohen because he rejects the labour theory of value.

These are, of course, very loose predictions, and Cohen would not pretend otherwise. In Popper's eyes they would fall far short of satisfying his desire for precise testability. There are no limits, no indications as to when relations of production will not be functional. This does not make the theory unfalsifiable, just very difficult to assess.[14] But methodologically it would be harsh to eliminate such a theory just because more precise prediction is available from smaller scale technological theories. If there were another theory of similar scope which produced more precise predictions and which stood up to the historical data better (perhaps Goldstone's demographic theory would be an example), then there would be a good case for rejecting the theory in favour of this alternative. But this is to place the theory on a par with other empirical theories, whereas the interest of Popper's anti-historicist arguments was that they seemed to allow us a way of criticising theories of history in advance of doing the empirical donkey-work. One has to come to a similar conclusion with regard to Goldstone's hypothesis. He has provided a great deal of empirical material which is arguably best interpreted in the manner he suggests. But again the predictive power of the theory is very weak. He asks whether revolution was inevitable (predictable) in the case of the English uprising, and suggests in reply that despite the set of circumstances predisposing the country to revolution, it just may have been possible for a very adept King to have avoided it. 'Perhaps Charles might

---

[14] See Joshua Cohen's review of *Karl Marx's Theory of History*, in *The Journal of Philosophy* vol. **79**, no. 5 (1982), for some counter-evidence.

have kept his head, and even his throne, if he had acted differently. But Charles was ruling in an explosive situation, and any number of false moves or accidents could have led to state breakdown' (*Revolutions and Rebellion*, p. 156). Despite the weakness of prediction, it does not seem plausible to fault Goldstone on methodological grounds. There is nothing *a priori* improper in his account.

Once the *ad hominem* moves against historicists are put aside, as we have suggested they must be, Popper is left with methodology, misery, and the supplementary knowledge argument. Methodologically, the type of 'large-scale' prediction which emanates from the theories of history of Cohen and Goldstone is tested by comparative histories of different societies and/or periods of history. It is true that this empirical check is less rigorous than the type of testing advocated by Popper—that of making small-scale adjustments to see what effects they have may well involve a greater degree of control over the various variables involved. However restricting one's investigations to this latter approach is also likely to lead to a far more cautious empirical sociology, one that avoids confronting questions such as those concerning the causes of revolutionary change. But if there are such causes, and if they form some kind of pattern of repeatable elements, then it would be cognitive cowardice if one were to avoid enquiring after them just because testing the proposed theories involved inevitable epistemological risk. Requiring that the sociologist be more rigorous may here be just a demand that the sociologist change the subject matter of their study.

We have not yet challenged the misery and knowledge arguments. Given that the knowledge argument concludes that prediction is impossible, it is probably best to consider that first, in the light of the preceding theories. The testing procedure envisaged by both Cohen and Goldstone is that of normal historical research, where causal and functional hypotheses are tested against the best data available.[15] This embraces generalising of the sort: if these factors were to be found elsewhere in similar (enough) circumstances, then we could expect the same types of effects to occur.[16] Where prediction of the future is impossible, then the best testing procedure involves retrodiction. Retrodiction is safe against the knowledge argument: there is nothing in the impossibility of knowing future knowledge that will tell against the possibility of the kind of retrodiction which is deployed by Goldstone in his testing of his *psi* hypothesis.

What of the predictions emanating from Marxism? Although retrodiction is deployed by Cohen and his critics (particularly in relation to the transition from feudalism to capitalism), here the knowledge argument threatens to be more powerful. This is because a prediction is central to the Marxist account of the dynamics of *capitalism*: capitalism will come to an end when the further development of the productive forces is inhibited by the (capitalist) relations of

---

[15] Functional hypotheses are particularly difficult to test, involving as they do questions about the selectionist history of the putatively functional traits. John Elder grapples with the problems in *Natural Selection in the Wild* (New Jersey: Princeton University Press, 1986).

[16] This ignores complications which arise on singularist accounts of causality, but as Popper was not such a singularist this is, in this context, justifiable.

production. Even ignoring the looseness of this prediction, we can see that the knowledge argument is relevant. How do we know what the future development of the productive forces will bring, given that future technology will embody future knowledge about the way gadgets work? The answer lies in asking what it is that we need to know about the productive forces in order to make this kind of prediction. The one thing that is absolutely essential is that future productive forces will replace present technology when, and only when, they are more productive, and so can produce more in a given amount of time.[17] The hypothesis is that our knowledge of the dynamics of capitalism licences us to predict that productive power will grow, and that eventually the specifically capitalist use (use geared towards profit) of this power will choke off further productive development, or will decrease the optimal rate of growth in such development, at which time the last capitalist crisis will occur. That is, as far as knowledge is concerned, we can know in advance what *type* of knowledge will become technologically embodied.[18] Future technology will enhance productivity,

[17] This whole account is a very crude presentation of much more complicated matters, but for present purposes a 'coarse grid' account will do. I am also not attending to any specific predictions emanating from the labour theory of value. For a more subtle account of the connections between social prediction, historical inevitability, determinism, and revolutionary agency, see Cohen's 'Historical Inevitability and Revolutionary Agency', in G. A. Cohen, *History, Labour, and Freedom* (Oxford University Press, 1988), pp. 51–82.

[18] A defence of social science prediction which emphasizes type of future knowledge is provided by Margaret Gilbert and Fred Berger, 'On An

and will embody that knowledge which helps it to do this. One could attempt to be slightly more definite (it would be difficult to be less specific!) and extrapolate from present trends, predicting that the revolution in computer/communications technology will continue, maintaining present trends toward computer controlled manufacturing and 'globalization'. One could then slot into this picture a more detailed economic and political forecast, depending on the favoured economic and political theories. The point is not that any of this will afford precise predictions. That is manifestly not so. What is at issue is whether the unknowability of future knowledge makes a theory of history impossible, and the argument is that given it is only a *type* of knowledge that is relevant, theoretical history is not ruled out.

4. Politics and Prediction

It may well be thought that none of the above makes the slightest dent in Popper's fundamental claim, which could be put as: predictions concerning consequences of social change are too vague to form the warrant supporting particular political actions, especially revolutionary action. Given such predictive vagueness the chances of creating unhappiness are too great. This makes it clear that the major weight of Popper's attack on historicism is borne by the misery argument. Despite the viability of theoretical history,

Argument for the Impossibility of Prediction in the Social Sciences', in *Studies in Epistemology*, N. Rescher (ed.) (Oxford: Blackwell, 1975).

large-scale political change is just too risky. The inbuilt imprecision of forecasts may be normal for theories of this scope, but what is justifiable for theorising fails to provide sufficient support for political action. The backing for this claim does not need to be philosophical: one does not have to show that detailed knowledge of the future is impossible in order to show that it is improbable. And one could even rest content with a lesser claim, that however things stand in the future, at present we do not have the historical know-how to justify revolutionary action. Such a pursuit of happiness can bring misery in its wake, and this misery is to be avoided.

There is much to be said for this stance. Looking at the history of revolutions does not give much support to advocates of the view that consequences of the overthrow of the state tend to be beneficial. Goldstone cites evidence of the human cost of revolution: in the English civil wars the ratio of the dead to survivors was 1 to 50; in the French revolution it was 1 to 20; in the Mexican revolution, 1 to 10. Moreover, the gain in political freedoms does not seem worth the risk either — 'history shows an almost uniform tendency of episodes of state breakdown to culminate in populist, usually military, dictatorship'.[19] The lesson for political actors seems clearly to be that advocated by Popper: acting so as to bring about revolutionary consequences is foolish, risking as it does the life and freedom of so many.

Political lessons are notoriously difficult to draw. One major problem is that one cannot simply compare a

[19] Goldstone *Revolutions and Rebellion*, p. 479.

present state of affairs with what *life was like* in order to assess what should have been done, or what political people should have advocated. The relevant comparison is with what life *would have been like now* had the revolutionaries not acted. Projecting oneself backward in time, it is not as though the political actor has only one prediction to make, the one concerned with consequences of revolutionary action. In order for the right comparison to be available a second prediction has to be made, that concerning the consequences of non-revolution. Given the relevance of this prediction, there are two fairly obvious objections to the sort of intermediary position held by Popper, a position which proposes neither revolutionary action nor quietism. One is that despite his stress on the unintended consequences of social action, Popper has implicitly a theory of revolutions which ties their existence (and consequences of their existence) too closely to revolutionary intentions. Goldstone notes that the failure of English institutions to cope with population exposition in the seventeenth century was not something *intended* by anybody. There were very few revolutionaries around. What was deemed desirable was change from those conditions which people were encountering. Goldstone notes that 'the key question about the seventeenth-century crisis in England is not what factors made people want change but rather what factors forced change onto people who were fundamentally conservative ...'[20] It looks like it was a series of 'small scale' changes that provoked the state

[20] Ibid. p. 155.

crisis, given the context in which those changes occurred (e.g. the inability of the institutions to adapt to the demographic transformation).

Second, and more damaging, is the reverse of the above objection. What the above example shows is that there can be revolution without revolutionaries because intention and consequence do not always go hand in hand. Similarly, there may be no small-scale change of the desired sort without the action of those advocating radical change. Many reforms are instituted in order to take pressure out of a brewing revolution. It is not unreasonable to suppose that people are inherently conservative in their social outlook, in the sense that there is a preference for a quiet life as opposed to one subject to change. In order to overcome this social inertia it may be necessary to threaten larger scale disruption in order to get any movement at all. Radical demands by racial and feminist groups may have paved the way for anti-discriminatory laws. The point here is that nothing in the Popperian critique of historicism tells against this possibility: for the small-scale change preferred by Popper, revolutionary demands may be needed.

My friends who rejected Popper's anti-revolutionary message did so in a more direct fashion than contained in the above considerations. Their context required a different response to that which would obtain in a democratic environment. However the irony is that one of the major pieces of policy to which they were opposed, the resettling of millions of black town dwellers in rural areas in South Africa, was indirectly responsible for the subsequent breakdown of the apartheid system. The resettling had the consequence of producing a huge population surge in rural areas, one that could not be contained and which eventually required the dismantling of the pass laws controlling the urban population. Even a Government desperate to retain its power can be the victim of the unintended revolutionary consequences of its conservative actions.

14   What Use is Popper to a Politician?

BRYAN MAGEE

Some years acquire symbolic status, and one such year is 1968. All over Europe and the United States university students exploded into violent rebellion. Insofar as this would-be revolution had an ideology it was unquestionably Marx-inspired, even if the Marxism was not always orthodox. It so happens that in the years 1970–1971 I was teaching philosophy at Balliol College, Oxford. And because of Oxford University's system, almost unique, of individual tuition for undergraduates, this meant I found myself in a continuing one-to-one relationship with bright students who were in the throes of revolutionary fervour.

Arguing with them was enormously illuminating for me. It seemed as if the more intelligent they were the more passionately Marxist they were—but also the more affected they were by intellectually serious criticisms of Marxism, which usually they were hearing for the first time. It was when they found themselves unable to meet these that they revealed where their fundamental motivation lay. This was not usually a positive one of belief in Marxist ideas. Still less was its commitment to communist forms of society, which usually they had been defending without knowing anything about the reality of them. The motivation was usually negative: it was inability or refusal to come to terms with their

411

own society as they saw it. Psychologically, this was nearly always at the root of their attitude.

Basically the chain of cause and effect between their ideas seemed to go something like this. They longed to live in a perfect society. But only too obviously the society in which they found themselves contained serious evils. So this form of society had to be rejected. A particularly interesting point here is the fact that, because what they demanded was perfection, they thought that if anything was seriously wrong then the whole must be rejected. If, say, newspapers reported cases of old and poor people dying of hypothermia in winter because they had no heating in their homes the students would say savagely 'There's something sick about a society that lets old people freeze to death in the winter'. If there were reports of students unable to take up university places because of an inability to get grants they would say 'There's something fundamentally rotten about a society that refuses to educate people unless they've got money.' It was virtually a formulaic response, of the fixed form: 'There's something fundamentally rotten about any society in which x happens', with x standing for any serious social evil. If anything *at all* was seriously wrong, the whole of society was sick: unless everything's perfect everything's rotten. Such an attitude could rest only on Utopian assumptions. And it quite naturally made those who held it receptive to a holistic as well as systematic social critique of the only society they knew. It also led most of them to suppose, erroneously, that there must be something somewhere that was infinitely better: since, plainly, things were not perfect here, they must be perfect somewhere else—or, at least, people somewhere else must be trying. Criticisms of communist reality were nearly always met by the counter-accusation that things were just as bad here, if not worse, and at least the Communists were striving to realize a moral ideal, which our cynical and self-interested politicians were not.

These attitudes display several errors of a fundamental character to which intelligent people in general are prone when they think about politics. Instead of starting from what actually exists, and trying to think how to improve it, they start from an ideal of the perfect society, a sort of blueprint in the mind, and then start thinking of how to change society to fit the blueprint. If they cannot see any practicable way of getting from reality to the blueprint they may be tempted then to think in terms of sweeping reality away, in order to start from scratch, in order to realize the blueprint.

Karl Popper's ideas are a marvellous antidote to such illusions. First of all he is insistent on its being an inescapable fact that wherever you want to go you have to start from where you are. Even the most cataclysmic revolution is an attempt to achieve certain ends, a way of trying to change society as it actually is into a different form of society that is preferred. And as the history of revolutions illustrates, existing society never is swept completely away: huge and important features of it always persist into the successor society, usually to the bafflement and chagrin of the revolutionaries. As a way of achieving desired social change revolution is exceedingly cost-ineffective as well as ineffectual. First and foremost, large numbers of people get killed, or are made to suffer appallingly in other ways. Second, desirable as well as undesirable social fabric is destroyed. Third, unrestrained violence on a large scale is uncontrollable when accompanied by a breakdown in the social order. Fourth, because it is uncontrollable the kind of society that emerges from it is nearly always one which the revolutionaries themselves say is quite different from what they wanted.

All forms of political thinking that start from blueprints of what is desired are anathema to Popper, and rightly so. All modern forms of society are in a state of perpetual change, and as time goes by the pace of this change gets faster, not slower. If we were to set ourselves the task of actualizing the most ideal blueprint, and then succeeded in actualizing it, even then change would not just suddenly stop. Marx and Engels thought it would—thought that with the realization of their perfect society history would come to an end. But nobody now believes this. Change will go on. So from the very moment we actualize our blueprint reality will start moving away from it and turning into something else. So the real political task is not to actualize an ideal state of affairs that can then be preserved for ever. This is the task to which the greatest political thinkers of the past, such as Plato and Marx, addressed themselves, but in reality it is not even an option. The real political task is to manage change.

As part of the process of perpetual change, peoples' aspirations and priorities perpetually change. So again, there too, even if we were able to start out with an ideal blueprint, and to succeed in our approach to it, as we worked towards it peoples' wishes would start moving away from it, so that even before we achieved it scarcely anybody would wholeheartedly want it. Something close to this has only too obviously happened in the late twentieth century with the ideal of socialism under its classic definition of public ownership and centralized planning of the means of production, distribution and exchange—an ideal which earlier in the century powerfully motivated millions of intelligent and well-meaning people, yet to which now scarcely anyone subscribes.

There is a need for perpetual revision of aspirations and goals, and this is inimical to the whole idea of a blueprint. Blueprints are fixed, static: if they changed unceasingly they would not be blueprints. They are therefore at best a source of never-ending problems, given the reality of permanent social change, and only too often they are a source of tragedy. Because they are fixed, peoples' attitudes towards them become fixed: they become objects of quasi-religious commitment and belief. And because they are seen as ideally desirable, political opponents who actively try to prevent them from being realized come to be looked on as wicked people who must be stopped, perhaps even removed from the scene altogether; and their elimination is seen as fully justified, indeed demanded, morally. Blueprints thus lead to rigidity, fanaticism, and through them to anti-rationality in many forms. The man with a blueprint usually knows he is right; and because of his utter certitude he feels justified in eliminating opposition by whatever means may be found necessary.

Popper's recommendation is that what we should eliminate are blueprints—eliminate them from our thinking entirely. Instead of basing our approach on an imaginary state of affairs that does not actually exist and is never going to exist, he recommends that we start from the social reality in which we find ourselves, and that we examine it critically to discover what is wrong with it, and to see how it may be improved. From that starting point he proposes what might be called a methodology for the management of change. I would like to go through this proposed method step by step.

First of all we are required to formulate our problems with care. That means, among other things, not taking for granted what they are. We have to ask ourselves what precisely are, say, the main problems that face us in the field of primary education? What, precisely, are the main problems that face us with the treatment of teen-age offenders against the law? What, precisely, are the main problems that face us in our relations with the United States? And so on and so forth.

There will, legitimately, be differences of opinion about what the problems are, before one has even begun to think in terms of solutions, and these differences should be thoroughly debated. It is of the utmost importance to get diagnosis right before one proceeds to cure, otherwise the proposed cure will be the wrong one, not effective, quite possibly harmful. So a lot of time and trouble and thought and work needs to go into the identification and formulation of problems before one attempts to move forward from that position.

Once a problem has been identified and clearly formulated, the next step is to consider alternative possible solutions. At this stage especially there can be opportunity for great boldness, and also for imagination and ingenuity, for freshness of perception and vision, for unexpected initiative. Usually it is here, if anywhere, that creative politics comes in.

But of course many if not most of the proposed solutions would not actually, if tried, work out very well in practice. As soon as you start to do something, anything, unexpected snags arise. Even in the most apparently sensible undertakings measures take longer than expected, or cost more, or prove to be administratively cumbersome, or alienate some of the individuals involved, or have unfortunate side-effects.

It is a matter of great practical concern that these drawbacks should be minimized by being foreseen and avoided. So proposed solutions need to be critically examined and debated, with the explicit object of bringing their faults to light before they are turned into reality. The more effective the criticism at this stage, the greater the saving in time, economic resources and human happiness, so a debate of this kind is not an abstract, airy-fairy matter, but a hard-headedly practical one. The proposals whose effective criticism is more desirable, because most fruitful, are those of government, because these are the ones that are put into practice on the largest scale, and with the most powerful backing, and with the greatest effect on peoples' lives. Full and free critical public discussion of proposed government policies is therefore essential if avoidable large-scale error is indeed to be avoided: without such discussion there will inevitably be more, and more costly, public-policy disasters than there need to be.

And of course there will be mistakes anyway. Even after a great deal of misplaced expectation has been eliminated by critical discussion, and the proposals thus critically improved are put into practice, things will still go wrong. Our actions have unforeseen consequences. So there is a need for practical as well as theoretical vigilance. After a policy has survived critical discussion and been put into practice, a critical eye needs to be kept on how it is actually working out, with a view to catching the first sign that is not working as hoped. At this stage the most important thing is not to be seeking reassurance that all is well, but the opposite, to be on the alert for the possibility that things are not going as they should. This requires the practical monitoring of public policy in action, and for that to be effective people need to be free to criticize not only a government's proposals but also its deeds. Again, the sooner harmful practices are identified the greater the saving will be in time, resources and human happiness. Governments that forbid public debate and criticism of their activities are bound to persist in mistaken, costly and harmful practices for much longer than they otherwise would; and being government activities these mistakes will usually be on a large scale.

It should always be remembered that the debate surrounding policies and their implementation may bring to light errors not only in them but also in the process one stage further back, the formulation of problems: we may come belatedly to see that our initial formation of our problem was wrong. Indeed, Popper remarks that we seldom really understand a problem fully until we have tried to solve it and failed.

This, in its barest outline, is the methodology rec-ommended by Popper to the practical politician. Some people may say it is embarrassingly obvious. I only wish it were. You do not need to be a very attentive reader of the serious press to realize that this is not how real-life politics is for the most part conducted. And as someone who was a professional politician for nearly 10 years I can assure you that the thought processes involved do not come easily to many politicians; indeed, some have serious difficulty in understanding them even when they are explained. If Popper's principles seem obvious to a philosophy-oriented audience it is because they are so rational, so congruent with situational logic. That is a powerful recommendation for them, but alas, it has not yet brought about their general acceptance or even comprehension. The task of actively promoting them still requires adherents.

Other critics may object that the whole approach is too cautious and therefore too slow. We haven't got time for all that talk, they may say: it's a luxury we can't afford. To this I believe the best reply is that of all possible political methods this is the one most likely to maximize the extent to which change remains under rational control. Attempts to short-circuit processes of criticism are almost bound to lead more error, and therefore more cost, and also more in the way of unintended consequences. There may indeed be more *change*, but disconcertingly much of it, too much, will not be in the required direction. This turned out to be one of the systematic shortcomings of centralized planning, and led in practice to its becoming almost invariably associated with systematized lying. Of course one cannot go on talking for ever. Decisions have to be made. But a debate that is genuine discussion and not just waffle or delaying tactics, although it may take time now, will save more than time later on.

The approach advocated by Popper is a broad recipe for effective and successful problem-solving. As such it has a general application to most practical affairs, not only to politics but to administration in any form, and also to business. People familiar with his philosophy of science and his more general theory of knowledge will have noted already that it instantiates his formula for problem-solving in those fields:

$$P_1 \rightarrow TS \rightarrow EE \rightarrow P_2$$

where P_1, is the initial problem, TS the trial solution proposed to this problem, EE the process of error elimination applied to the trial solution, and P_2 the new situation thus arrived at, with its new and sometimes unexpected problems. In fact the relationship between Popper's methodology of politics and his theory of knowledge is so close that it is worth our going on now to look at some specific features that they have in common.

First, Popper regards himself in both fields as addressing not a static or stable state of affairs but a process of change, and he sees the main challenge as being how to manage change, in one case the growth of knowledge, in the other ongoing social development. In both cases he sees the demands this makes on us as consisting above all else of problem-solving. In both cases, therefore, he thinks we should start from the careful analysis and understanding of problems, and not leap straight away to what is in fact the second stage, the proposal of trial solutions.

In politics solutions, real or attempted, are normally called policies. Every reputable political or social policy is a proposed solution to a problem; and we always need to be clear about the problem before we can propose the solution. We must always be able to ask of a policy: 'To what problem is this the solution?' If there is no problem to which a given policy is a solution then the policy is superfluous, and therefore harmful, if only because it consumes resources to no purpose. Policies which are not solutions to any identifiable problem are part of the common currency of so-called practical affairs. Committees are especially good at producing them. I have stopped many a committee meeting dead in its tracks by asking the question: 'To what problem is this the solution?' The whole notion that you can start with policies is deeply erroneous, and very damaging in practice. One of the forms it takes is starting from a blueprint, because of course a blueprint is a proposed solution; but it takes many other and more mundane forms. It is essential to start from *problems*, and to arrive at the formulation of each policy only as a solution to a problem.

According to Popper, in both politics and the growth of knowledge, criticism is the most effective agent of desirable change, and must therefore be not only free but welcomed, and acted upon. We can never be in a position to know that we have got things right; our formulations and policies are always open to improvement; therefore any notions of certainty or unquestionable authority are not only out of place but damaging. The best we can *do*, like the best of our knowledge, is the best only for the time being, and in the prevailing circumstances. It is always, in principle, improvable, and therefore should always be subject to critical discussion.

In practice this attitude ought to breed a respect for political opponents, and a willingness to learn from them. In all the democracies I know, politicians lag behind the public on this matter. They would be more, not less, popular with their electorates if they were more willing than they are to admit error, and they would also be more, not less, popular if they were more willing than they are to admit that their opponents are quite often right.

The Popper approach constitutes a programme for practical and rational improvement, and the usual word for that in politics is 'reform': so it is a methodology of reform. But it leaves open the question of how quick or slow reform should be, the even more important question of how radical it should be, and the most important question of all, namely what it should consist of. This makes it an approach that can be adopted by anyone on the political spectrum between those who want no change at all and those who want revolution. What this means in practice is that it can be adopted by anyone committed to democratic politics: so it is also what you might call a methodology for democracy. It so happens that the youngish Karl Popper who wrote *The Open Society and Its Enemies* in the late 1930s and early 1940s had always been left of centre, and throughout the whole of his adult life up to that point a strongly, emotionally committed social democrat. But like so many people he moved to the right in middle age, and by the time of his death would have been accounted a conservative by most people—though to the end of his days he continued to regard himself as a liberal in the classic sense of the word, meaning someone who puts individual liberty first among the political values. My point is that his basic approach is one that can be adopted by anyone committed to democratic politics, from the extreme democratic left to the extreme democratic right, which indeed was the gamut that Popper himself passed through.

Having said that, though, the point has to be made that the Popperian approach sits most comfortably with a left-of-centre position, the sort of position Popper himself occupied when he produced it. This is because it gives rise naturally to a radical attitude towards institutions. It is not only policies that have to be seen as attempts to solve problems: institutions do too. A country's education system is its solution to the problem of how to educate its young; its armed forces are its solution to the problem of how to defend itself; its health services are its solution to the problem of what public provision to make for those of its citizens who need medical help; and so on and so forth. Just as in the case of policies, an institution that is not a solution to any problem is superfluous—indeed, it is that condition that renders institutions obsolete. And because an institution is a practical solution to a problem, so long as it has a real function it is capable of being more effective or less, more satisfactory or less, more comprehensive or less, more expensive or less, more popular or less, and so on. The Popperian approach involves subjecting institutions to a permanently critical evaluation in order to monitor how well they are solving the problems they exist to solve—and involves moreover a permanent willingness to change them in the light of changing requirements. I have always taken the famous dictum of Jesus of Nazareth 'The Sabbath was made for man, and not man for the Sabbath' to mean that we should bend institutions to fit human beings, not human beings to fit institutions; but this is at odds, I do believe, with some of the basic attitudes common to political conservatism, which include a reverence for institutions as such, a deep-seated unwillingness to change them, and a readiness rather to let their requirements override personal considerations. There is no logical incompatibility, but there is, I think, a certain psychological uncomfortableness in combining a Popperian approach to the requirements of institutional change with a typically conservative emotional attachment to existing institutions. The only kind of conservative with whom the two can sit comfortably together are those of the radical right, politicians like Margaret Thatcher, whose approach to traditional institutions was in fact highly disruptive.

The permanent monitoring of institutions to see if they are *not* performing as required, and the permanent monitoring of the implementation of policies to see if they are having undesirable consequences, are activities—and reflect a cast of mind—that come much more readily to radicals, of left and right, than they do to traditional conservatives. They also run counter to the way people working in institutions, especially those with authority, tend normally to behave. The normal tendency is to cover up organizational and administrative failures as much as possible, and to resist facing even to oneself the fact that one's activities are not having the desired effects. The Popperian approach, which requires one actively to seek out failures and short-comings and do something about them, calls for a degree of intellectual honesty from politicians and administrators, as it does from scientists, that does not come to them at all easily, and constitutes a disconcerting personal challenge. What provides the incentive to meet this challenge is the higher success rate that results from doing so.

In fact a thoroughgoingly problem-solving approach has many practical advantages, perhaps even more in politics than in science. It is far easier to get agreement on problems than on solutions, and a government that starts from the problem—let us say, to take a small but emotive example, the problem of what to do about the number of homeless people sleeping rough on the streets of London—and then shows itself open to alternative possible solutions will prob-ably have not only a higher degree of practical success than one that starts with the answer, in other words a policy; it will also enjoy more support and goodwill, even from those who disagree with what it eventually does. In a democracy a great deal of electoral advantage is to be had from a problem-solving approach, because people will feel that they have been brought in.

And of course, if I may be forgiven for stating the obvious, a problem-solving approach directs one's attention to problems, and makes doing something about them the first priority. It protects one from being seduced into trying to build Utopia; and yet it does not easily allow one to relapse into complacency or inactivity. One's energies are channelled not into constructing ideal models but into removing avoidable evils. Popper encapsulates the first rule of thumb he recommends for public policy in the words 'Minimize avoidable suffering.' Psychologically it is a different approach from that of crusading for ideals, to which so many political activists are dedicated: it is more practical, and nearly always more fruitful. In any case the two are not necessarily incompatible. I am not opposed to idealists as such, but I do regard them with the gravest of suspicion. It is a fact that social evils have been perpetrated by idealists in our century on a simply stupendous scale that includes the deliberate murder of tens of millions of men and women and the herding of tens of millions more into forced labour camps (I am thinking not only of the Soviet Union but also of China, where the numbers involved may have been greater). These things could not possibly have been done by people who had adopted 'Minimize avoidable suffering' as their guiding principle. But they were done by idealists, and condoned all over the world by other idealists, more often than not with a sense of moral self-righteousness accompanied by savage denunciations of anyone who criticized what they were defending.

A point Popper makes which I stress more than he does is the unavoidability of unintended consequences. I stress them because they often dominate practical politics —as they soon came to do in all communist societies, for example. An awareness of them also immunizes us against enthusiasm for any form of centralization, especially centralized planning. To anyone engaged in practical affairs, business as well as politics, they are of never-ceasing importance. Only on someone divorced from reality can they fail to impinge.

Political lessons to be learnt from Popper are not confined to the problem-solving approach and its method. He has certain large-scale perceptions about politics that seem to me right and important although unfashionable. For instance, he perceives clearly that the societies in which we in the West are living in the 1900s are by all real (as against ideal) standards—that is to say by all the standards of past experience—exceptionally non-violent, as is the international scene as a whole. He also sees that for the great majority of men and women in the democratic West life is better now than it has ever been before, not only materially but in the most important non-material ways, for example health, education, and cultural opportunity. He therefore sees clearly that the cultural pessimism so fashionable today, when intellectuals and artists are saying on all sides that we live in a uniquely terrible and violent time, presents more or less the opposite of the truth. I suspect that the illusion it represents has been brought about partially by the collapse of the historicist, progressivist illusions that were held earlier in the century by a great many of the same people, and to which Popper was equally opposed. On the face of it, it is peculiar that so many individuals who for decades believed with a kind of religious intensity that everything was getting better are now equally certain that everything is getting worse. But both attitudes are holistic and uncritical, and meet what seem to me primarily religious emotional needs. The fact is that the liberal democracies of the West are the only large societies in the whole of human history in which the great majority of the people have enjoyed not only material prosperity and literacy but also what have come to be known as fundamental human rights. This is a very recent historical phenomenon, and it is a wonderful thing. Even so, there is no contradiction at all between seeing this clearly for what it is and at the same time trying to improve these societies, and for that purpose adopting a radical and essentially critical stance in their political and social affairs. It happens to be the position I myself have always occupied, independently of Popper, and it is what first drew me to his work, before I knew anything about his epistemology or his philosophy of science.

Another overall perception of Popper's which I share is that equality of outcomes is not a desirable social goal. It took me a long time to learn this lesson, and when I did it was not from Popper but from my poor constituents in East London. They were almost entirely without social envy, which I came through them to realize is a largely middle class phenomenon anyway. They wanted a better deal for themselves—better wages, better houses, better schools for their children, and so on—but had no desire to pull down anyone who was better off. On the contrary, they actively rejected any such attitude; it ran counter to some of their most basic aspirations, more often for their children than for themselves. And they saw it as incompatible with elementary personal freedom. They were right in this. And it was also Popper's view. He once said that if a form of socialism could have been discovered which was compatible with personal freedom he would still be a socialist.

Another general attitude of Popper's that I loudly applaud in his hostility to the tyranny of fashion in all its forms—the idea that we have to do certain things, or do things in certain ways, because these are the 1990s, and that we really have no choice, in that anything else is contrary to the spirit of the times, and therefore inappropriate, perhaps even inauthentic. This error is at its most predominant and destructive in the world of the arts, but it operates in politics too. In Britain after World War Two we had years of uncritical commitment to Keynesian economic management followed by uncritical commitment to monetarism; we had an uncritical belief in nationalization followed by an uncritical belief in privatization. Town planners guided by what they took to be the spirit of the times devastated the centres of many of Britain's most beautiful towns during the 1960s and 1970s, and corralled the poor of the inner cities into tower blocks. Anyone who opposed these developments at the time was denounced as conservative or reactionary, fuddy-duddy, out of date. Popper has always believed in either fighting or ignoring such tides of opinion. He sees them as forms of what another kind of philosopher would call 'false consciousness', and as ways of evading responsibility for our own decisions and our own actions. Insofar as we go along with them we are enemies of our own freedom. We can do *whatever* we can do, and it is up to us to do the best we can.

One of Popper's specific proposals that I think has great merit is that it should be accepted internationally as a fundamental principle that no existing frontier is to be changed except by peaceful negotiation. The point here is that nearly all the national frontiers in the world were established by force, usually either imposed on the vanquished by the victors in war or imposed on colonized peoples by imperialist powers; therefore if the fact that a frontier has been imposed without the consent of one of the parties is to be accepted as an excuse for that party to use violence to get it changed there would be justified wars breaking out all over the world all the time, several on each continent. This cannot be acceptable to the international community now. Existing frontiers, constituting as they do actually existing political reality, must be regarded by the United Nations as operative no matter how they were arrived at, and must be guaranteed by whatever international peace-keeping forces there are, unless a majority of those whose frontiers they are wish to change them by peaceful means.

Up to this point I have been endorsing Popper's approach and commending it to you. And the truth is I do believe it provides working politicians with rules of thumb of the utmost usefulness. But it does have, inevitably, limitations and shortcomings. The chief limitation is that, being a methodology, it is almost entirely about method and not about content. The most pressing question facing the individuals who have to take important decisions is nearly always 'What should we do now?' Everyone else can stand back from that question and then criticize the way things are done, but the decision-makers themselves cannot. Only rarely does the Popperian approach help them towards an answer. This fact has recently come to the fore in the former communist countries of Eastern Europe and the Soviet Union. To an extent rare in history they have found themselves with opportunities to build a new society that is radically different from the one they had before. Popper's philosophy offers them first-rate guidance about how to do things, but very little about what to do. What kind of local government, if any, do they want: at what level, how constituted, and with what powers? What kind of education system do they want, what sort of schools, how organized, by whom, teaching what? How much welfare state do they want, and in what areas—and how much can they actually afford: how is it to be administered, how funded? It is questions like these that constitute most of the content of large-scale practical politics.

In any case, most politics is not large-scale. When I became a Member of Parliament and began spending my days in the House of Commons among hundreds of other MPs, I was struck by the fact that, among themselves, they scarcely ever discussed the sort of political or social questions thrashed out in pubs and debating societies, like are we in favour of the return of the death penalty, or censorship, or nationalization. The questions that held them in thrall were much more like: 'If we raise the widow's pension by half a percentage point where are we going to find those extra millions of pounds?' They would have differing views about such questions, and would argue heatedly, but these mostly were the sorts of questions they would be arguing about. And it is inevitable that these are the sorts of questions that day-to-day government has to concern itself with. It is seldom that Popper's work offers much guidance with them.

This in itself is not a criticism of Popper, because he is not talking to us on that level. From a philosopher a politician must expect strategic, not pragmatic, guidance. What I am drawing attention to is not a shortcoming but a limitation. It is, however, one that practical politicians are likely to be a lot more conscious of than other people.

Practical politicians are only for a very small part of the time concerned with putting principles into practice. Most of the time they are struggling to make the best they can of difficult, messy and uncontrollable situations. I will give you an example of this that involves a conflict between me and Popper personally. I have already mentioned his conviction that the international community should impose an iron refusal to allow existing frontiers to be changed by force, and have given his reasons for it. Well, when the military junta then governing Argentina invaded the Falkland Islands, for which Britain was responsible in international law, and *de facto* war began, he telephoned me at the House of Commons in great passion, wanting me to urge the British Government to declare war formally on Argentina. I refused. What I said to him went roughly as follows: 'I agree that the Argentinians absolutely must be made to leave, by negotiation if possible but by force if necessary. And I will vote for the use of force if there is no other way. But I want to get them out with the minimum possible harm to everyone concerned, and I see this as a damage-limitation exercise. There happens to be a sizeable British community living permanently in Argentina that consists of tens of thousands of families, many of whom have been there since the nineteenth century. They have their own schools and other institutions, as well as their own homes, businesses and professional practices. If we declare war on Argentina, the Argentinian government may well intern them and confiscate their assets. Their whole world will be destroyed, and in many cases their individual lives will be ruined. I believe we can get the Argentinians out of the Falkland Islands without that happening—though only if we don't declare war.'

Popper, always willing to sacrifice himself to a principle, was willing to sacrifice others too, and would not agree with me. Not only did he continue to telephone me angrily throughout the Falklands war, always urging the same course of action on me; he continued to bring the subject up with me for the rest of his life, always maintaining that he had been right. I am convinced to this day he was wrong—and not only because what I wanted to happen did in fact happen. I fully acknowledge that it might not have done. But I am convinced that we were right to try. I re-emphasize that I was always completely in agreement with Popper that in no circumstances should Argentina be allowed to get away with the forcible annexation of the Falkland Islands. He and I differed only about how they were to be made to leave. But on this we differed profoundly. It was not the principle that was in dispute but the way it should be put into practice. Popper wanted commitment to the principle to be publicly proclaimed in a formal act: I saw this as unnecessary to the actual implementation of the principle and almost bound to be seriously damaging. So I saw my own approach as essentially practical and his as essentially theoretical—but far too theoretical, culpably so, too little concerned with the actual lives of individual men, women and children. And I have to say, as an intellectual and academic myself, that I see this fault as all-pervading in the attitudes of intellectuals and academics to political and social matters, and as being an extremely serious, often debilitating fault. Also, having been a professional politician as well, I find the sense of personal superiority to politicians so commonly expressed by intellectuals and academics unfounded and misplaced, self-deluding.

This story of a clash between a political philosopher and a professional politician illustrates a point of profoundest importance. I do not believe that there are many people who hold Popper and his work in higher regard than I do; and I knew him well personally. As a professional politician I made conscious use of his methodology, and found it of extraordinary practical usefulness and fruitfulness. Yet any individual who, if only by his vote in an assembly, has to take responsibility for executive political decisions, is likely to find himself unable to put Popper's principles—or anybody else's principles, for that matter—into practice in a way that the originator of the principles would wholly approve of. This is because practice has unavoidable and compelling exigencies which theory can never encompass, and which those who are solely theoreticians seem only rarely to appreciate—and never fully to understand. But that would be a subject for a different paper.

WORKS OF KARL POPPER REFERRED
TO IN THE TEXT

Abbreviation	Title, etc.
AWP	*A World of Propensities* (Bristol: Thoemmes Antiquarian Books Ltd., 1990)
BG	*Die Beiden Grundprobleme der Erkenntnistheorie* (originally written 1930–2) (J. C. B. Mohr, 1979)
'BNIQM'	'Birkhoff and von Neumann's Interpretation of Quantum Mechanics', *Nature*, **219** (1968), pp. 682–5
CR	*Conjectures and Refutations* (London: Routledge and Kegan Paul, 1963, 4th ed., 1972)
'IA'	'Intellectual Autobiography', in P. Schilpp (ed.), *The Philosophy of Karl Popper*, 2 vols. (La Salle, Illinois: Open Court, 1974), pp. 1–181. (see also *UQ*)
'IINE'	'Indeterminism is not Enough', *Encounter* **40** (April 1973), pp. 20–26. (Reprinted in *OU*)
IQP	'Indeterminism in Quantum Physics and in Classical Physics', *British Journal for the Philosophy of Science*, **I**, (1950), pp. 117–33, 173–95
KBMP	'Knowledge and the Body-Mind Problem: In Defence of Interaction*, M. A. Notturno (ed.) (London and New York: Routledge, 1994)
'KU'	Zur Kritik der Ungenauigkeits relationen', *Die Naturwissenschaften*, **22** (1934), pp. 807–8
LdF	*Logic der Forschung* (Vienna: Julius Springer Verlag, 1934)
LSD	*The Logic of Scientific Discovery* (translation of *LdF*) (London: Hutchinson, 1959; 3rd ed., 1972)
MF	*The Myth of the Framework: In Defence of Science and Rationality*, M. A. Notturno (ed.) (London and New York: Routledge, 1994)

(*cont.*)

Abbreviation	Title, etc.
'NSD'	'Normal Science and Its Dangers', in I. Lakatos and A. Musgrave (eds), *Criticism and the Growth of Knowledge* (Cambridge University Press, 1970), pp. 51–58
'NSEM'	'Natural Selection and the Emergence of Mind', *Dialectica* **32** (1978), pp. 339–355
'OCC'	'Of Clouds and Clocks', (Washington University, St. Louis Missouri, 1966), reprinted in *OK*
OK	*Objective Knowledge* (Oxford: Clarendon Press, 1972; 2nd ed., 1979)
OS	*The Open Society and Its Enemies* 2 vols. (London: Routledge and Kegan Paul, 1945, 5th ed., 1966)
OU	*The Open Universe* (London: Hutchinson, 1982). Vol. 2 of the Postscript to *LSD*
'PDR'	'Remarks on the Problems of Demarcation and of Rationality' in I. Lakatos and A. Musgrove (eds), *Problems in the Philosophy of Science*, (North Holland, Amsterdam, 1968), pp. 88–192
PH	*The Poverty of Historicism* (London: Routledge and Kegan Paul, 1957)
'PIP'	'The Propensity Interpretation of Probability', *British Journal for the Philosophy of Science*, **10** (1959), pp. 25–42
'PMN'	'The Place of Mind in Nature', in Richard Q. Elvee (ed.), *Mind in Nature* (San Francisco: Harper and Row, 1982), pp. 31–59
'QM'	'Quantum Mechanics without "The Observer"', in M. Burge (ed.), *Quantum Theory and Reality*, (Springer: Berlin, 1967), pp. 7–44
QTSP	*Quantum Theory and the Schism in Physics* (London: Hutchinson, 1982). Vol. 3 of the Postscript to *LSD*
RAS	*Realism and the Aim of Science* (London: Hutchinson, 1983). Vol. 1 of the Postscript to *LSD*
'RC'	'Replies to my Critics' in P. Schilpp (ed.), *The Philosophy of Karl Popper*, 2 vols. (La Salle, Illinois: Open Count, 1974), pp. 959–1197

(cont.)

Abbreviation	Title, etc.
'RSR'	'The Rationality of Scientific Revolutions', in R. Harre (ed.), *Problems of Scientific Revolution* (Oxford: Clarendon Press, 1975), pp. 72–101
SB	(with J. Eccles) *The Self and Its Brain* (Berlin: Springer International, 1977)
UQ	*Unended Quest* (London: Fontana Collins, 1976). A reprint, with some changes, of 'IA'

INDEX

438

Printed in the United States
by Baker & Taylor Publisher Services